SpringerWienNewYork

Hans-Martin Henning (Ed.)

Solar-Assisted Air-Conditioning in Buildings

A Handbook for Planners

Second Revised Edition

SpringerWienNewYork

Dr. Hans-Martin Henning
Fraunhofer Institute for Solar Energy Systems ISE, Freiburg, Germany

Printed in Austria
Springer-Verlag Wien New York is a part of
Springer Science + Business Media
springer.com

Coverphoto: PSE GmbH, Solar Info Center, Freiburg, Germany
Layout: Tim Selke, Freiburg, Germany
Printing: Druckerei Theiss GmbH, A-9431 St. Stefan, Austria

Printed on acid-free and chlorine-free bleached paper
SPIN: 12077809

With 126 Figures

CIP data applied for

ISBN 978-3-211-73095-9 Springer-Verlag Wien New York
ISBN 3-211-00647-8 1 st edition Springer-Verlag Wien New York

This handbook is dedicated to our colleague Cees Machielsen (†),
an expert on absorption systems and one of the European pioneers in the investigation of
solar-assisted cooling and refrigeration and above all, a good friend.

PREFACE

Interest in the use of solar thermal energy for air-conditioning of buildings is growing continuously among planners and other professionals. The motivation for this is the wish to achieve comfortable indoor conditions in buildings by using environmentally friendly technology. This handbook aims to contribute towards achieving this goal by supporting professionals during a project's design phase and providing information on both suitable components and systems. Therefore, the main focus lies on technologies and equipment which are commercially available today or which are at least at the stage of pilot testing. The key components covered are thermally driven cooling systems - both closed cycles producing chilled water and open cycles directly treating ventilation air - and solar collectors as the major heat source to drive the cooling system.

Today, about 50 solar-assisted air-conditioning systems are installed in the countries participating in Task 25 'Solar-Assisted Air-Conditioning of Buildings' of the International Energy Agency (IEA) Solar Heating and Cooling Programme. This underlines the fact that this technology is still in an early stage of development. Almost no standardised design guidelines exist and there is still a lack regarding common practices for design and construction. Field data and experience gained from installations under real operating conditions has shown that there are frequent shortcomings in the system's hydraulic design, as well as with the controls. Furthermore, in some cases, the expected energy savings could not be achieved in practice.

Some of these weaknesses can be avoided by following the information and guidance provided in this handbook. However, it is very important to note that a solar-assisted air-conditioning system requires a greater effort during the design phase than a conventional system for the same application. Often, it will be necessary to perform annual computer simulations of several different system configurations in order to identify the one with the best energy-cost performance. In addition, based on today's experience, it is highly recommended to keep the hydraulic design as simple as possible. Then a comprehensible operation strategy and a transparent control scheme can be implemented, reducing the risk of error or malfunction. An appropriate commissioning process for the entire installation will also have to be implemented given the greater complexity of systems using solar energy compared to conventional systems. Finally, field monitoring of operating conditions and performance, e.g., web-based or by means of telecommunications networks, is strongly recommended to allow troubleshooting and prompt identification of component malfunctions or control failures. Although this is valid for any modern large HVAC installation, it is even more essential in the case of solar-assisted air-conditioning systems.

The main goals of this handbook are to encourage planners and potential users to consider the installation of solar-assisted air-conditioning systems and to provide them with helpful information during the decision-making and the design processes. A properly designed and carefully operated installation will create a high degree of satisfaction by providing a high level of indoor comfort to the users while using environmentally friendly technologies.

EDITOR'S NOTES

This handbook is product of a co-operative work carried out by experts from eleven countries in Task 25 'Solar-Assisted Air-Conditioning of Buildings' of the Solar Heating & Cooling Programme of the International Energy Agency (IEA). All contributing authors are grateful to the national funding authorities which enabled work within the Task, as well as support for the production of this handbook.

Thanks are due to all Task participants and co-authors who followed the iterative process of writing and reviewing this handbook, a process which included extensive discussions and sometimes many, time-consuming iteration loops. Of all the persons who contributed to the production of this book, some should be named personally. Particular thanks are due to Constantinos Balaras who made a very comprehensive review of the entire book, with many, very helpful suggestions. A great job was carried out by Mario Motta, who compiled all the text several times, included suggestions, re-arranged text sections, re-structured the whole book and finally succeeded in bringing it into a very consistent form. Helen Rose Wilson not only provided the special support of a native speaker but also served as a perfect test candidate, with an ideal mixture of basic technical understanding but without too deep involvement in the specific topics of the handbook. David Marold from the publisher Springer was not only very flexible with regard to timing but also provided us with many valuable tips on the transfer of technical information to the target audiences. Last, but not least, Tim Selke spent many days, nights and weekends to bring all the material into the right form and format for attractive and reader-friendly presentation.

Hans-Martin Henning, Editor

AUTHORS, CONTRIBUTORS

In addition to the Task 25 participants in general, the following individuals authored and contributed significantly to this handbook. The appropriate chapter and section numbers follow each contributor's name; their coordinates can be found in Appendix 4:

Jan Albers (5)
Constantinos A. Balaras (8.1, review of entire handbook)
Christoph Kren (3.3.1)
Marco Beccali (3.3.2)
Gerdi Breembroek (3.1)
Maria Joaó Cavalhó (4.1.3)
Uwe Franzke (3.1, 3.2, 6.6, Appendix 1)
Gershon Grossman (3.2.4)
Hans-Martin Henning (1, 3, 4, 5, 6, 7, 8.3, review of entire handbook)
Carsten Hindenburg (4.1.4, 8.3)
João Farinha Mendes (4.1.3)
Rodolphe Morlot (3.3.3, 4.1, 4.2, 6.2)
Mario Motta (2, 3, 5, review of entire handbook)
Daniel Mugnier (2.2.1, 3.3.3, 4.2, 6.2)
Jean-Yves Quinette (3.1.1, 6.2, 8.2)
Rien Rolloos (2, 3.1.3, 3.2.2)
Tim Selke (Appendix 1, review of entire handbook)
Wolfgang Streicher (4.1, 4.2, Appendix 1)

1 INTRODUCTION

Buildings represent one of the dominating energy-consuming sectors in industrialised societies. In Europe about 40% of primary energy consumption is due to services in buildings /1.1/. Both, private and commercial buildings consume energy for applications such as heating, hot water, air-conditioning, lighting and other - mainly electrically operated - equipment.

AIR-CONDITIONING AND ENERGY CONSUMPTION

Figure 1.1
Flat-plate collector field for air-conditioning of a factory building in Inofita Viotias/ Greece.

During the last few decades the energy consumption for air-conditioning purposes increased dramatically in most industrialised countries, even in heating-dominated climates. In 1996 about 11,000 GWh primary energy were consumed in Europe alone by small room air-conditioners up to a cooling capacity of 12 kW. According to EU studies, this value is expected to increase by a factor of 4 to about 44,000 GWh by 2020 /1.2/. This figure does not include the centralised air-conditioning plants or chilled water systems that are generally installed in large commercial buildings. The main reasons for the increasing energy demand for summer air-conditioning are the increased thermal loads, increased living standards and occupant comfort demands, and architectural characteristics and trends, such as an increasing ratio of transparent to opaque areas in the building envelope to even the popular glass buildings.

LOAD REDUCTION AND PASSIVE CONCEPTS

Figure 1.2
Evacuated tube collector field for air-conditioning of a wine cellar in Banyuls/ France.

Energy conservation is an approach to reduce the disadvantages of the continuously increasing energy demand for air-conditioning in both an economic and an environmental sense. Nowadays, the know-how on building design concepts leading to energy load reduction is broadly based and well developed. It is well established in practice based on results from extensive research studies, with technology aiming to reduce cooling energy needs, for example, external shading devices, improved daylighting concepts in combination with intelligent control of artificial lighting and energy-saving equipment.

A further step has been taken towards the use of 'cheap' cooling sources, for example, heat sinks such as the outdoor air for night cooling, or evaporative cooling, radiative cooling, ground cooling using earth-to-air heat exchangers. However, in general the cooling capacity of these passive (or natural) cooling techniques is rather limited and cannot fulfil the cooling requirements of all building types, with different end user needs and climatic conditions.

SOLAR THERMAL COLLECTOR SYSTEMS

In addition, during the last thirty years and particularly in the past decade, growing environmental concerns and consistent effort in research and product development have opened up the market for active solar systems. Significant advances and improvements have been made, resulting in a wealth of experience with large installations using solar thermal collectors. An example of a large, flat-plate collector field for solar air-conditioning is shown in Figure 1.1 and an example with evacuated tube collectors in Figure 1.2. In spite of a significant and growing market penetration rate, the main obstacle preventing broad application of solar thermal collectors beyond their use in domestic hot water production has been the seasonal mismatch between heating demand and solar energy gains. Long-term storage units have to be employed in order to overcome this.

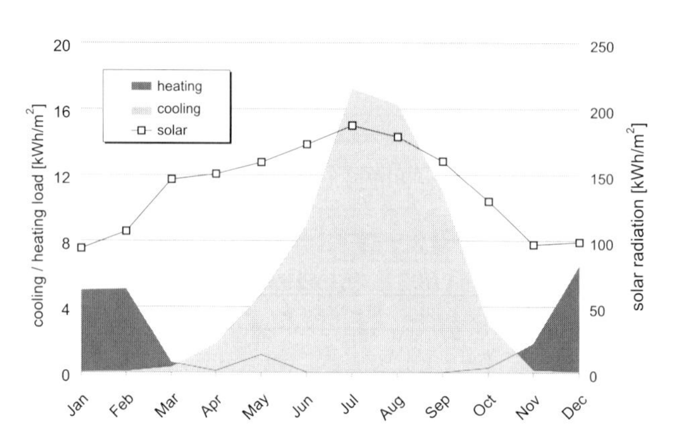

Figure 1.3
Example of cooling and heating loads (kWh per m² room area per month) and available solar radiation (kWh per m² collector area per month) for a site in southern Europe.

The need for seasonal storage does not arise if solar thermal energy can be exploited for air-conditioning of buildings during summer, i.e., sensible cooling and air dehumidification. The great advantage for this kind of application is that the seasonal cooling loads coincide with high solar radiation availability. An example of the annual distribution of monthly heating and cooling loads and the available solar radiation is shown in Figure 1.3. The main goal of applying solar energy for air-conditioning is to reduce the energy consumption and to make building air-conditioning more environmentally friendly. The exploitation of solar energy throughout the year, for both heating and cooling, improves the performance output of a solar thermal installation and the economics of the investment.

WHY USE SOLAR ENERGY FOR AIR-CONDITIONING?

The first demonstration of solar-assisted absorption cooling machine was made during the Paris World Exhibition in 1878 by Augustin Mouchot, based on a technique developed by Edmond Carré /1.3/. Following intensive research and development activities in this field in the United States and Japan mainly during the 1980s (see e.g., /1.4/, /1.5/), there was a slow-down during the 1990s. However, recently this topic has attracted new interest in many countries. The reasons for this are manifold. On the one hand, there is an increased consciousness of the environmental problems which are created by the use of fossil fuels for generating electricity consumed by conventional cooling systems. In addition, the use of common working fluids (refrigerants), with their ozone-depleting and/or global warming potential, has become a serious environmental problem. Also, the impact of air-conditioning in increasing the peak power demand and contributing to shortages in the electricity supply have caused serious problems in several countries. This underlines the need to implement advanced, new concepts in building air-conditioning.

On the other hand, the current conditions for the application of solar technology are much more favourable than in the past decades. Industrial production of solar components - solar thermal collectors as well as photovoltaic arrays - is well established in many countries. Reliable products with improved efficiency and competitive costs are available, which guarantee a reliable energy supply for many years without disproportionate maintenance costs. In addition, automated control technology has advanced and improved greatly due to new developments in semi-conductors, which makes it feasible to handle even complex systems and installations efficiently. Subsidy programmes for application of solar technology in many countries document the political intention to increase the use of renewable energy sources as a substitute for fossil fuel consumption.

PHYSICAL PRINCIPLES

In principle, there are many different ways to convert solar energy into cooling or air-conditioning processes; an overview is given in Figure 1.4. A main distinction can be made between thermally and electrically operated systems. Among the thermally driven processes, thermo-mechanical processes and processes based on heat transformation can be distinguished. The latter are all based on reversible thermo-chemical reactions with relatively low binding energies. Open cycles are in contact with the atmosphere and always use water as the 'refrigerant'. Closed cycles can also use other refrigerants, e.g., ammonia.

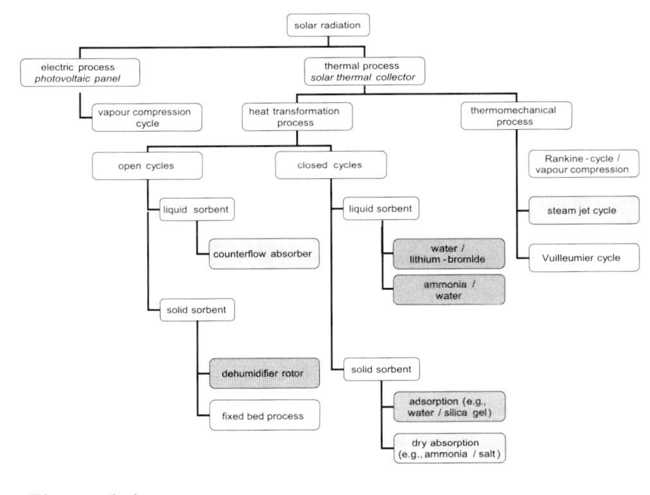

Figure 1.4
Overview on physical methods to use solar radiation for cooling or air-conditioning. Processes marked in dark grey: commercially available technology which is used for solar-assisted air-conditioning. Processes marked in light grey: technology at the stage of pilot projects or system testing.

SOLAR ELECTRIC OR SOLAR THERMAL AIR-CONDITIONING?

Solar energy can be converted directly into electricity using photovoltaic panels, to drive a vapour compression chiller with an electric motor. However, in industrialised countries, which have a well-developed electricity grid, the maximum use of photovoltaics is achieved by feeding the produced electricity into the public grid. From an economic point of view, there is an even greater incentive if the price for electricity generated by solar energy is higher than that of electricity from conventional sources (e.g., feed-in laws in Germany or Spain). Therefore photovoltaically powered solar air-conditioning systems are of minor interest from a systems point of view, unless there is an off-grid application.

The competition between different types of solar technology, i.e., photovoltaic versus solar thermal systems, in the case of air-conditioning applications raises the question: is it more appropriate to install solar-thermally driven air-conditioning or to install a grid-connected PV power system and use a conventional air-conditioning system? In the course of a building project, this question has to be answered considering both energy performance and economic issues.

TECHNICAL SCOPE

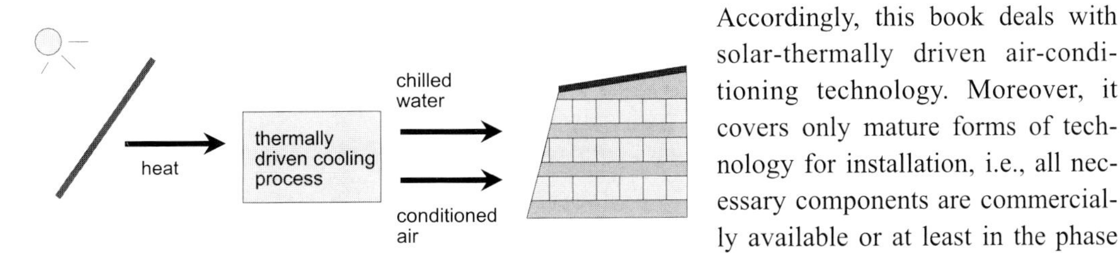

Figure 1.5
General scheme of systems covered in this handbook.

Accordingly, this book deals with solar-thermally driven air-conditioning technology. Moreover, it covers only mature forms of technology for installation, i.e., all necessary components are commercially available or at least in the phase of pilot applications. This also justifies the emphasis on centralised systems that provide conditioned air and/or chilled water to an entire building or certain areas of a building, since small (2 - 4 kW) thermally driven systems for decentralised application in a single room are not yet available on the market. The general type of systems covered in this handbook is demonstrated in Figure 1.5.

This handbook does not address issues on how to reduce the building loads for air-conditioning or energy conservation; many textbooks and handbooks on this subjects are available (see e.g., /1.6/). The system boundaries considered here are the incoming solar radiation on one side and the chilled water and/or conditioned air that is supplied to the building on the other side.

Presented Technology - Water Chillers

The dominating type of thermally driven cooling technology to produce chilled water is absorption cooling. Absorption chillers have been in commercial use for many years, mainly in combination with cogeneration plants, using waste heat or district heating. For air-conditioning applications, absorption systems commonly use the water/lithium bromide working pair. Another closed-cycle sorption technology to produce chilled water uses the physical process of adsorption but this kind of chiller has a much lower market share. Nevertheless, there are many installations that use solar-thermally driven adsorption chillers.

Presented Technology - Desiccant Cooling

Another type of technology which has gained increased attention over the last 15 years is desiccant cooling technology. Using this technology, air is conditioned directly, i.e., cooled and dehumidified. Desiccant cooling systems exploit the potential of sorption materials, such as silica gel, for air dehumidification. In an open cooling cycle, this dehumidification effect is generally used for two purposes: to control the humidity of the ventilation air in air-handling units and - if possible - to reduce the supply temperature of ventilation air by evaporative cooling.

Target Audience

The goal of this handbook is to support planners in designing a solar-assisted air-conditioning system which uses solar energy as a heat source and a thermally driven cooling / air-conditioning system to meet the load. Typical questions, which the book intends to answer, are:

- Is the use of solar-assisted air-conditioning feasible for a given building at a specific site?
- Which technology can be used?
- Which is the best system for the given application under the conditions of the specific site?
- Which solar collector types should be used for the selected air-conditioning system?
- What dimensions of the solar collector area and other system components result in the best energy cost-performance?
- Which tools are available that can support a user to further address these questions?

Handbook Structure

This book consists of two main parts: part I (Chapters 2 to 4) covers components of solar-assisted air-conditioning systems and part II the entire systems (Chapters 5 to 8).

Part I - Components

In Chapter 2 a classification of air-conditioning systems is made and their main technical data are described and summarised.

In Chapter 3 the main types of technology to transform heat into cold are described, i.e., thermally driven chillers and desiccant cycles. Standard components such as vapour compression chillers and cooling towers are also addressed briefly, since they may play an important role in solar-assisted air-conditioning either as part of the system or as a reference to which a solar driven system has to be compared.

In Chapter 4 solar collectors, the main driving energy source in solar-assisted air-conditioning, are described. Hot water storages, being an important part of solar heat production systems, are also discussed, as well as other heat sources which may serve as a back-up.

PART II - SYSTEMS

In Chapter 5 a step is taken from the component level to the system level. Different complete systems are compared and general operation strategies are discussed. The main systems categories are air systems, water systems and combined air-water systems.

In Chapter 6 the design of complete systems is presented, whereby the main focus is placed on the design of the solar system in combination with the air-conditioning system. Diverse design approaches are presented which provide different levels of detail, depending on the available information.

In Chapter 7 performance figures are defined, which allow the comparison of different technical solutions for solar-assisted air-conditioning with each other and the comparison with conventional systems that are used as a reference. A key figure to evaluate a system is defined which combines economic and energy-related aspects of performance.

In Chapter 8 three design examples are presented that provide an overview of relevant technical and financial data, and overall performance.

PART I - COMPONENTS

2 THE LOAD SUB-SYSTEM - AIR-CONDITIONING EQUIPMENT

Air-conditioning, or indoor climate control, is the process of treating air so as to control simultaneously its temperature, humidity, cleanliness and distribution. The designer is responsible for considering various systems and recommending the one that will best provide the desired air treatment. It is imperative that the designer defines the goals of the design according to the user needs and project constraints.

BUILDING LOADS

Before starting any planning activity for air-conditioning systems - for solar-driven systems in particular - it is always important, to make sure that the estimated building loads are reasonable. It is always more economical and rational to apply energy-saving measures - at almost any planning stage - to achieve reasonable loads instead of overdimensioning the plant in order to meet excessive loads. Once the main design constraints, such as use of the building, cooling loads, occupancy schedule, required ventilation rates and zoning requirements, are clear, the air-conditioning design process can be initiated. In some cases, for example low-energy buildings, this will be the first step of an iterative process, which will lead to the optimisation of the building envelope and installations. Given the project constraints, the designer will have to determine the feasible technology options that will fulfil the user requirements.

CLASSIFICATION OF AIR-CONDITIONING SYSTEMS

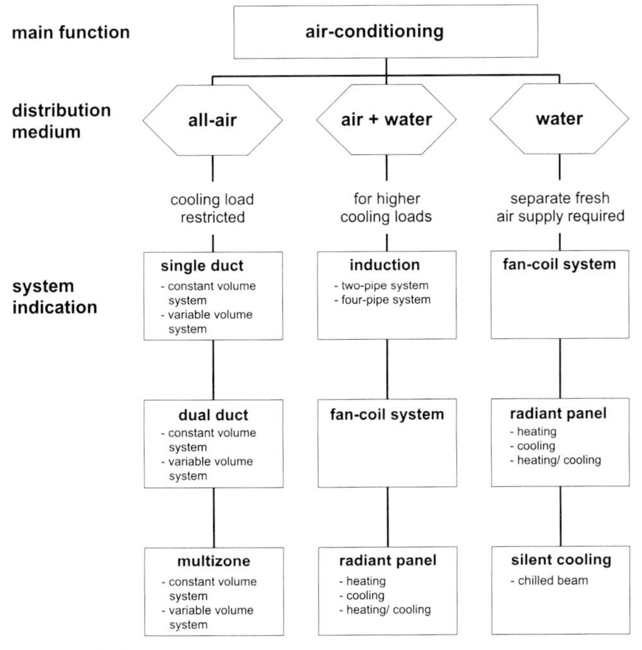

Figure 2.1
Generic classification of centralised air-conditioning systems.

Air-conditioning systems fall into one of four major categories: all-air, air-water, water and refrigerant-based systems. Each type has certain technical and economic advantages. Some are better than others for specific applications. This identification is based on the controllable fluids used in each zone for the appropriate control of the air conditions. Each system or combination of systems has specific capabilities concerning cooling capacity.

The refrigerant-based systems have a distinct feature compared to the other three groups. They are normally unitary systems, meaning that the 'cooling' process takes place at a very short distance from the delivery terminals; these systems are mainly window- or wall-mounted.

In the other, centralised systems, the refrigeration units may be located at some distance from the conditioned space, for instance in a central mechanical equipment room. The distribution system for the cold medium (supply and return) connects them to the delivery terminals. Only centralised solar-assisted air-conditioning systems are currently available on the market, although there are research efforts towards unitary systems. For the purpose of this book, only centralised systems will be considered further. The generic classification for centralised systems is shown in Figure 2.1. Table 2.1 lists specific cooling capacities of the different air-conditioning systems.

	Air-Conditioning system	Ventilation rate	Room height in m $\Delta T = T_{room} - T_{supply}$ (draught-free conditions)				
	System type	air change per hour [h^{-1}]	2.4 T = 6°C	2.7 8°C	3.0 10°C	3.5 12°C	4.0 15°C
A	All-air systems	3	15	20	30	40	60
"		4	20	30	40	55	80
"		5	25	35	50	70	100
"		6*	30	45	60	85	120
B	Displacement ventilation	approx. 8	30	30	30	**	**
C	Induction/ fan-coil	10	50	75	100	140	200
D	Chilled ceiling	-	60	60	60	60	60
E	Chilled floor	-	20	20	20	20	20
F	Pure displacement	250	800	900	1	1.15	1.35
A	+ D	2	70	75	80	90	100
"	"	3	75	80	90	100	120
"	"	4	80	90	100	115	140
"	"	6	85	100	100	140	180
B	+ D	approx. 8	80	80	80	**	**
C	+ D	10	90	100	100	140	200

Table 2.1

Specific (maximum possible) cooling capacities [W/m²] of different air-conditioning systems /2.1/. For cooling loads higher than 45 W/m² choose: air cooling based on minimum required fresh air quantity (e.g., 30 to 50 m³/h per person) and secondary cooling (water-based); e.g,. system C or A + D. The aim is to save costs for ducting and energy for the transportation of air and to avoid possible draughts when introducing too much air into the room. Possible draught can be avoided by using, for example, high-induction air outlets in the wall or swirl outlets in a ceiling.

**) At higher ventilation rates*

***) Values not known.*

2.1 All-air systems

An all-air system provides complete sensible and latent cooling, and possibly humidification of the supplied air. No additional cooling or dehumidification is required in the zone. The same air stream is normally used for heating purposes. Air is delivered to the conditioned space through air ducts and distributed through air outlets (i.e., grilles, slot diffusers, ceiling diffusers, perforated ceiling panels, variable diffusers etc.) or mixing terminal outlets (mixing boxes). Indoor air is then usually returned to the main air-handling unit, where heat/energy is recovered, or exhausted to the outdoors.

All-air systems may be adapted to many applications for comfort or process work. They are often used in buildings that require individual control of multiple zones, such as office buildings, schools, hospitals, laboratories and hotels.

Advantages and Disadvantages

The advantages of all-air systems are that they have the greatest potential for use of outside air for 'free' cooling, and they provide a wide choice of zoning, flexibility and humidity control under all operating conditions, with possible simultaneous heating and cooling, even during off-season periods. Heat-recovery systems can be easily incorporated into the main air-handling units, thus resulting in considerable energy savings. By adjusting the volume air-flow, they are able to maintain practically constant indoor space conditions within a very small tolerance.

Their disadvantages include the requirement for additional duct space within the building for air distribution and their higher installation and operation costs. Their installation and design require closer co-operation between architectural, mechanical and structural designers.

All-air distribution systems do not pose specific problems if applied for solar-assisted air-conditioning, and no components need to be specially adapted. Therefore the reader can refer to existing specific texts for more thorough treatment (e.g., /2.2/).

Height of the room in m	ΔT in °C
2.4	6
2.7	8
3	10
3.5	12
4	15

Table 2.2
Maximum temperature difference for supply of cooled air.

COMFORT

The supply-air temperature and speed delivered to indoor spaces have to be carefully controlled in order to avoid thermal discomfort and draught problems. The maximum possible temperature difference between the supply and indoor air depends on the height of the indoor space, actually on the height of the mixing zone above the occupation zone (1.8 m above the floor level). Furthermore, the way of introducing the air influences the temperature difference. From practical experience (at least in northern European countries), it has been documented that air-cooling complaints related to draught are minimised when the temperature difference between exhaust air and supply air, ΔT, is not greater than the values listed in Table 2.2. Typical capacity values for such systems are shown in Table 2.1.

All-air systems can be classified into five groups: single-duct systems in the versions variable air volume (VAV), or constant air volume (CAV), single-duct displacement systems, dual-duct systems and multizone systems.

In general, single-duct systems consume less energy than dual-duct systems, and VAV systems are more energy-efficient than CAV systems. Energy waste due to leakage in ducting and terminal devices can be considerable, reaching values up to 20 % /2.2/. Heat-recovery devices can save considerable amounts of energy and reduce the required capacity of primary cooling and heating.

A brief overview of the most important all-air type systems is presented in the following sections.

2.1.1 Single-duct systems: variable air volume (VAV) and constant air volume (CAV) system

This system supplies a single stream of conditioned air (either warm or cold air) from the air-handling unit to the indoor space. Single-duct CAV systems change the supply air temperature and humidity in response to the space load, while maintaining a constant air-flow. The VAV system controls the indoor air temperature by varying the quantity of the supplied air rather than varying the supply air temperature. The supply air temperature is held relatively constant, depending on the season. The greatest energy savings associated with VAV systems occur in the building's perimeter zones, where variations in solar load and outdoor temperature permit a reduction of the supplied air quantities.

2.1.2 Single-duct displacement systems

This system supplies the conditioned air near floor level at a low speed. For cooling, the air-supply temperature is some degrees below the indoor air temperature. The supplied air spreads out over the floor. Indoor air movement is enhanced - from bottom to top - because of convective heat gains (i.e., from occupants, equipment, etc.). Cooling capacity is restricted to about 30 W/m^2 at a room height of 2.7 m. Air movement may also be enhanced by the removal of the exhaust indoor air.

In clean rooms and similar zones, pure displacement ventilation can be applied with high ventilation rates and high specific cooling capacities.

2.1.3 Dual-duct systems

This system supplies two streams of air at different conditions (one warm air stream and one cold air stream) from the air-handling unit to the indoor spaces. The two ducts run in parallel and deliver air to a mixing box. The warm and cold air streams are mixed in correct proportions to satisfy the specific space loads at each conditioned space or zone.

Dual-duct systems use more energy than single-duct VAV systems and require even more space for the ducts. Note that these systems may be designed as constant air volume or variable air volume systems.

2.1.4 Multizone systems

This system supplies several zones from a single, centrally located air-handling unit. Different zone requirements are met by mixing cold and warm air through zone dampers at the central air-handling unit in response to zone thermostats. The mixed, conditioned air is distributed throughout the building by a network of single-zone ducts. This is a common system in humid climates; it often produces air at a temperature of 9 - 11°C for humidity control. Supply air is then mixed with a return stream to maintain comfort conditions in different zones. Heat is only added if the zone served cannot be maintained by delivery of return air.

2.2 Water systems

Water systems heat and/or cool a space by direct heat transfer between water and the indoor air. They provide means to cover indoor cooling (or heating) loads but cannot satisfy the ventilation loads. Therefore, fresh air has to be provided by other means, for example either by infiltration in naturally ventilated buildings or by other local fresh/exhaust air exchange or even a central, mechanical ventilation system. Depending on the type of chilled-water air-conditioning terminal unit installed in the space, it may be possible to handle latent loads. The most common terminal units that are used with chilled-water systems include fan-coils, chilled ceilings, chilled floors or systems using natural convection in combination with cooling coils (silent cooling, chilled beams). A brief overview of the most common terminal units is presented in the following sections.

2.2.1 Fan-coils

Basic Description

A fan-coil system is a heat exchanger (coil) with a fan that simply circulates indoor air over it, housed in the same unit. The heat exchanger is supplied with hot or chilled water. The units have a thermostatically controlled built-in fan that draws air from the room and then blows it over finned tubes of the heat exchangers where hot water or steam for heating or chilled water for cooling is circulated. The hot or cold medium is centrally produced either from equipment located in the building or supplied by a district heating/cooling network.

A sketch of a simple fan-coil unit is shown in Figure 2.2. The system can be controlled by using a simple thermostat or a more complex electronic microprocessor control system (three-way valves, variable speed fan, etc.). Fan-coils are examples of single-zone systems. They are available in horizontal ceiling-mounted, concealed, or recessed vertical floor units.

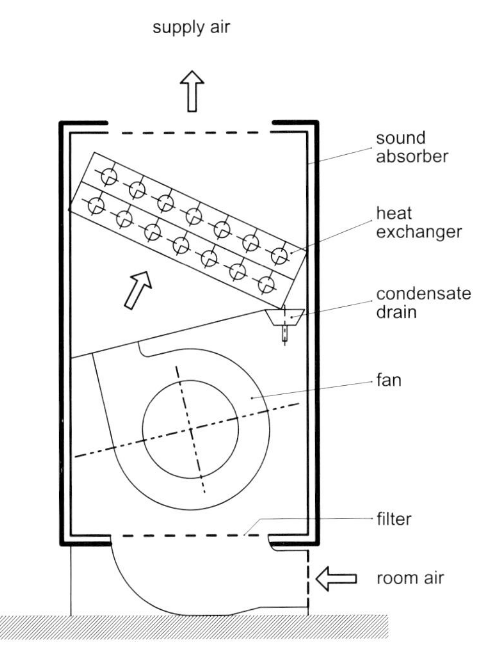

Figure 2.2
Cross-section of a typical simple fan-coil unit with one heat exchanger for air heating/cooling.

Types of Fan-Coils

Two types of fan-coils are commonly available, namely two- or four-pipe systems.

- Two-pipe system: A two pipe unit uses one pipe for the supply and one pipe for the return of the hot/cold medium to the heat exchanger. The unit can then be used for either heating or cooling the indoor air, depending on the central system operation mode. Multiple chillers and/or boilers may be required for multiple zones. Unit-mounted controls utilise a pipe-mounted sensor; it is used to determine the operation mode, i.e., heating or cooling, and the unit is controlled accordingly.
- Four-pipe-system: A four-pipe unit is equipped with two independent coils, one for heating and one for cooling. Cooling and heating valves for controlling coil capacities are often factory-installed, and their control devices are hidden inside the unit's cabinet, or they are wall-mounted, or remotely mounted.

Standards

Beside these systems, also two-pipe systems with an electric heater are available. Standard operating conditions for the use of fan-coils are for instance defined in Eurovent Standard 6/3 /2.3/. Accordingly, in the cooling mode, the standard defines the water supply/return temperatures at 7/12°C for an indoor air temperature of 27°C (relative humidity 50 %). In the heating mode, the standard defines the water supply/return temperature at 70/60°C for a 4-pipe system and 50/40°C for a 2-pipe system, for an indoor air temperature of 20°C.

Advantages and Disadvantages

The main advantages of fan-coils are:

- The system requires only piping installation, which takes up less space than air ducts in all-air systems.
- Unoccupied building spaces may be separated by simply turning off the local fan-coils or diverting the cold/hot medium flow to the fan-coil.
- Zones can be individually controlled and managed with a centralised control unit.

The main disadvantages are:

- Condensate must be removed from each unit.
- Interior zones may require additional fresh-air ventilation with separate ducts.
- Heat-recovery may be more difficult to achieve.
- Potentially noisy system since the air fans are located inside occupied areas.

Energy Performance

For the purpose of this handbook it is of particular interest to consider the energy performance of typical fan-coil units. Based on information from commercially available fan-coils, typical cooling capacities of a single fan-coil can vary between 0.5 kW and 10 kW. In general, the heating capacity is around 2 times higher than the cooling capacity. The typical air-flow is in the range of 5.5 to 6.5 m^3/h per kW of cooling capacity. A fan-coil can be operated in a range of around 10% to 100% of the full cooling capacity by controlling the air-flow. The electric consumption for the fan is in the range of 3% to 7% of the cooling capacity if the fan-coil runs at the lowest air-flow and in the range of 1% to 2.5% of the cooling capacity if the fan runs at its maximum. Most fan-coils are equipped with a device to dispose condensate in those cases where control of indoor humidity is possible. Condensation occurs, depending on indoor conditions, when the air is cooled below its dew-point.

2.2.2 Chilled ceilings

Characteristics of Chilled Ceilings

A cooling system which is based on chilled ceilings and ventilation separates latent and sensible cooling. The main part of the sensible cooling is delivered to the room by the chilled ceiling. Cooling is transferred to the room via radiation and/or convection. The relative proportion of each heat transfer mechanism depends on the design of the chilled ceiling. A ceiling with a closed surface involves primarily radiation for the heat transfer and is referred to as a radiative ceiling, whereas a convective ceiling presents a more 'open' structure. A closed loop circulates water through the ceiling. The water inlet temperature is usually between 16°C to 18°C. Changing the water flowrate and/or the water inlet temperature controls the room temperature. Dew-point sensors may be installed to avoid condensation by increasing the inlet water temperature or reducing the water flowrate.

A chilled-ceiling system is composed of a network of tubes in which cold water circulates; generally the tubes are either made of plastic materials or metal (usually copper). The network of tubes is either attached to the ceiling and covered by a false ceiling, or directly attached to the elements of false ceilings, or clipped onto metal diffusers.

Chilled ceilings are available in two different generic types:

- For drop-in ceiling applications (panel with back insulation and acoustic perforation).
- For free-hanging designs (design for enhanced upper surface).

Cooling Capacity

A major disadvantage of chilled ceilings is that no latent loads can be covered. Also their cooling capacity is rather limited. Typical values are in the range of 70 W/m^2 for drop-in ceiling applications and up to 140 W/m^2 for free-hanging designs (both figures are valid for a difference between room air temperature and average water temperature of 10°C). For chilled ceilings, typical time constants are in the range of three to five minutes, as result of the lightweight construction of the panels (approximately 5 kg/m^2).

2.2.3 Chilled floors

Basic Description

The cooling/heating floor system consists of a network of tubes, in which hot or cold water circulates, depending on the seasonal needs. The floor network is often made entirely of plastic materials. It is then placed on an insulated layer (to reduce ground heat losses) and the entire construction is embedded in a floating slab.

Floor heating systems are generally a well-established technology and common practice in residential or large commercial buildings. There is a European Standard on how to design and dimension floor heating systems /2.4/. On the other hand, floor cooling is a relative new idea, although several successful applications have demonstrated their applicability to maintain comfort conditions in a variety of buildings.

Advantages and Disadvantages

The main advantage of a chilled floor is that the device is completely integrated into the floor and thus is perfectly invisible and does not diminish the exploitable indoor space. Due to the inertia of the floor, a mismatch between the supplied cooling power and cooling loads is buffered by the system; in fact, care should be taken that the floating slab (including the floor covering) does not have a very high thermal inertia (mass above the top of the insulation limited to about 160 kg/m^2). Another advantage is that the system operates silently since no air is moved actively. From an energy point of view, the main advantage is that the temperature of the chilled water supplied to the floor ranges between 12°C - 20°C (depending on the load) and is considerably higher than for fan-coil systems, where they typically range between 7°C - 12°C. Similar advantages occur during the heating mode, when the floor water-supply temperature is considerably lower than for conventional systems.

The main disadvantages are the limited capacity and that it is not possible to cover latent loads. Depending on local weather conditions, appropriate control of the chilled-water supply temperature and flowrate to the floor system is very important in order to avoid condensation problems. The use of appropriate ventilation to enhance indoor air circulation is essential for optimising space cooling.

Cooling Capacity

The cooling capacity of a typical system lies in the range of 35 - 40 W/m^2 (indoor temperature 26°C, floor surface temperature 20°C).

2.3 Air-water systems

Classification

Air-water systems condition spaces by distributing air and water to terminal units installed in the zones to be conditioned throughout a building. The air and water are cooled or heated in central equipment rooms and from there are distributed to the air-conditioned spaces. The following systems can be distinguished:

- Induction system (two-pipe or four-pipe).
- Fan-coil system with supplementary air.
- Radiant panels with supplementary air.

Air-water systems apply primarily to perimeter building zones/spaces with high sensible loads. They may, however, be applied to interior zones/spaces as well. These systems work well in office buildings, hospitals, hotels, research laboratories and other buildings where their functions meet the performance criteria.

2.3.1 Induction systems

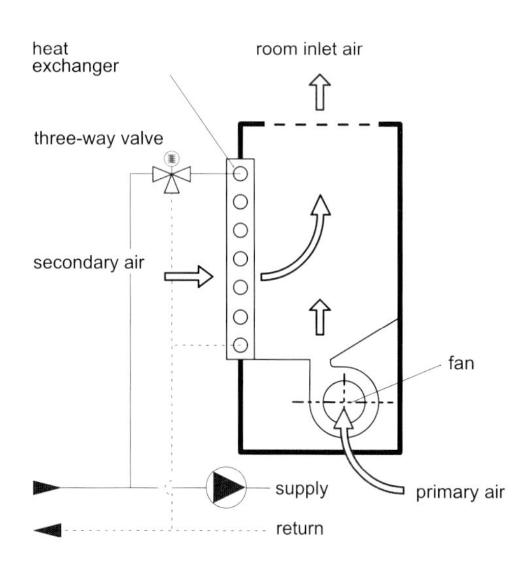

Figure 2.3
Schematic drawing of a two-pipe induction unit; primary air comes from the central air-handling unit, secondary air is indoor air which is cooled and blown back into the room.

These systems use a high speed, high-pressure and constant air volume supply to a high-induction type terminal unit. High pressure means, that narrow ducting is required. Induced air from the room is either heated or cooled within the terminal as required. The capacity control is implemented by means of water flow regulation or a bypass. This system may use two pipes (one water circuit; supply and return) or four pipes (two independent water circuits) for heating and cooling. The primary air (fresh air) is usually conditioned in a central air-handling unit, while the secondary air (indoor air) is conditioned in the terminal unit. Therefore, this system requires central production of chilled (hot) water, that is supplied to the air-handling unit and the local terminal units. A schematic drawing of a two-pipe induction unit is shown in Figure 2.3. Induction units are usually installed in building perimeter zones/spaces, for example, under a window, although units designed for overhead plenum installation are gaining popularity.

2.3.2 Fan-coil systems with supplementary air

This type of system is a combination of a fan-coil with a centralised air-distribution system to provide fresh air. The advantage in comparison to a simple fan-coil is that heat-recovery (heating season) and indirect evaporative cooling (summer) of ventilation air can be easily achieved. A simpler alternative is to have a wall opening where the fan-coil draws outdoor fresh air. This applies to perimeter building zones only.

2.3.3 Chilled building components with supplementary air

The combination of chilled building components, for example, ceilings, floors or walls, with a central ventilation system can provide high-comfort air-conditioning. The conditioned fresh air is supplied using a central air-handling unit. Sensible loads not covered by the ventilation air are extracted by the system of chilled building components.

3 THE COLD PRODUCTION SUB-SYSTEM

Thermally driven chiller-based and desiccant systems are key solutions for solar-assisted air-conditioning systems. Therefore, this section deals with the most common types of thermally driven chillers, namely absorption chillers and adsorption chillers, and in particular the ones which are feasible for coupling with a solar thermal energy source. Desiccant cooling technology is also presented and a brief description of the processes and components employed is given. Furthermore, vapour compression chillers are introduced since they may be used as a cold back-up source in a solar air-conditioning system and they serve as a reference for comparison between solar-assisted and conventional systems.

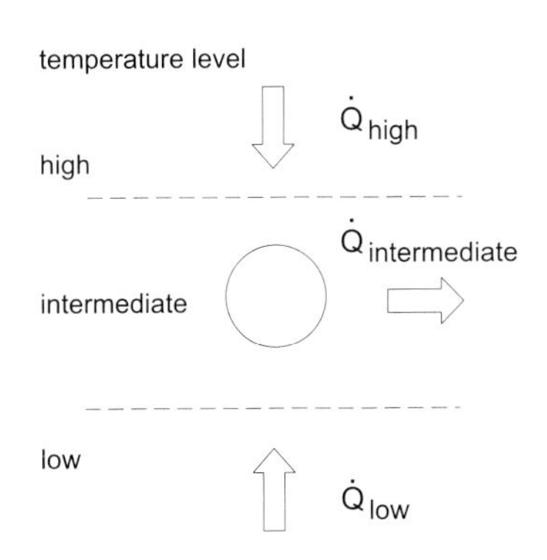

Figure 3.1
Schematic diagram of energy flows in a thermally driven machine operating a refrigeration cycle.

PROCESS PRINCIPLE

A refrigeration machine consumes energy to transfer heat from a source at a low temperature, to a sink at a higher temperature. In case of air-conditioning the heat extracted from the low temperature source is the useful cooling, i.e., the heat removed from the conditioned space thereby producing the cooling effect. In the vast majority of air-conditioning applications, the intermediate temperature heat sink is the external environment and the heat is rejected to the external air. The driving energy is heat in case of a thermally driven process and is mechanical energy in case of a conventional refrigeration machine. In most cases the mechanical energy is delivered by an electrically driven motor, at least in case of building air-conditioning.

As a result of the first law of thermodynamics, the flux of heat rejected at the intermediate temperature level, $\dot{Q}_{intermediate}$, is equal to the sum of the heat flux extracted from the low-temperature heat source, $\dot{Q}_{low}$, and the driving power of the process, P_{drive}, i.e., $\dot{Q}_{intermediate} = \dot{Q}_{low} + P_{drive}$. For an electrically driven chiller, the driving power is the electricity input to the motor, P_{el}. In the case of thermally driven chillers, the driving energy flux, P_{drive}, is a heat flux at a high temperature level, $\dot{Q}_{high}$. Figure 3.1 shows the principle for the example of a thermally driven chiller.

COEFFICIENT OF PERFORMANCE

A key figure to characterise the energy performance of a refrigeration machine is the *Coefficient of Performance*, COP. For thermally driven air-conditioning systems, the $COP_{thermal}$, which indicates the required heat input for the cold production, can be defined as follows:

$$COP_{thermal} = \frac{\dot{Q}_{low}}{\dot{Q}_{high}} = \frac{\text{Heat flux extracted at low temperature level}}{\text{Driving heat flux supplied to cooling equipment}} \quad (3.1)$$

The $COP_{thermal}$ varies with the equipment operation conditions, i.e., the three temperature levels, the percentage of load etc.; therefore COP-values of different systems are only comparable if the same operation conditions are considered. For a conventional, electrically driven vapour compression chiller, the COP_{conv} is defined as the required electricity input for production of cooling energy:

$$COP_{conv} = \frac{\dot{Q}_{low}}{P_{el}} = \frac{\text{Heat flux extracted at low temperature level}}{\text{Electrical power supplied to the chiller}} \quad (3.2)$$

The COP-values of conventional chillers and thermally driven refrigeration machines cannot be directly compared since the quality of the energy input (exergy content) is different. A method that is commonly used for an appropriate comparison is based on the primary energy consumption. This method is elaborated in Chapter 7 of this handbook.

3.1 Chillers

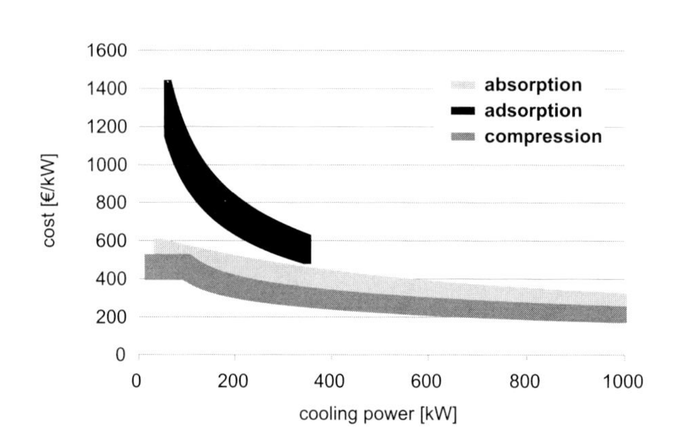

Figure 3.2
Specific cost ranges for different chiller types as a function of the cooling power (cost figures include heat rejection device, i.e., air cooling or cooling tower).

In building air-conditioning chillers are used to produce chilled water which is supplied to the air-conditioning components as described in Chapter 2. For the combination with solar thermal collectors thermally driven chillers are of particular interest. Main types used in solar air-conditioning systems are absorption chillers which use a liquid sorbent and adsorption chillers which use a solid sorbent. Vapour compression chillers are also described here, since they serve as the conventional reference and - in some cases - are used as back-up for chilled water production.

Chiller Cost

The cost of a chiller strongly depends on the capacity of the machine. As example a range of typical costs for chiller-based systems in €/kW is given in Figure 3.2; these values are based on a market survey in Central Europe /3.1/.

3.1.1 Absorption chillers

The working principle of an absorption system is similar to that of a mechanical compression system with respect to the key system components evaporator and condenser. A vapourising liquid extracts heat at a low temperature (cold production). The vapour is compressed to a higher pressure and condenses at a higher temperature (heat rejection). The compression of the vapour is accomplished by means of a thermally driven 'compressor' consisting of the two main components absorber and generator. Subsequently, the pressure of the liquid is reduced by expansion through a throttle valve, and the cycle is repeated.

Absorption Cycle

Absorption cycles are based on the fact that the boiling point of a mixture is higher than the corresponding boiling point of a pure liquid. A more detailed description of the absorption cycle includes the following steps:

1. The refrigerant evaporates in the evaporator, thereby extracting heat from a low-temperature heat source. This results in the useful cooling effect.
2. The refrigerant vapour flows from the evaporator to the absorber, where it is absorbed in a concentrated solution. Latent heat of condensation and mixing heat must be extracted by a cooling medium, so the absorber is usually water-cooled using a cooling tower to keep the process going.
3. The diluted solution is pumped to the components connected to the driving heat

source (i.e., generator or desorber), where it is heated above its boiling temperature, so that refrigerant vapour is released at high pressure. The concentrated solution flows back to the absorber.

4. The desorbed refrigerant condenses in the condenser, whereby heat is rejected at an intermediate temperature level. The condenser is usually water-cooled using a cooling tower to reject the 'waste heat'.

5. The pressure of the refrigerant condensate is reduced and the refrigerant flows to the evaporator through an expansion valve.

A schematic drawing of a basic absorption cycle is shown in Figure 3.3.

WORKING PAIRS

The heat required for step 3 can be supplied, for instance, by direct combustion of fossil fuels, by waste heat or solar collectors. Depending on the required cooling effect, one of the following working pairs for absorption chillers is commonly employed:

- For a temperature of the low temperature heat source higher than 5°C, for example when used for air-conditioning a water/lithium bromide (LiBr) pair absorption machine, is most frequently used, which must be water-cooled.
- For a temperature of the low temperature heat source lower than 5°C, for example, when used for refrigeration, an ammonia/water machine can be used, which can be cooled by either air or water.

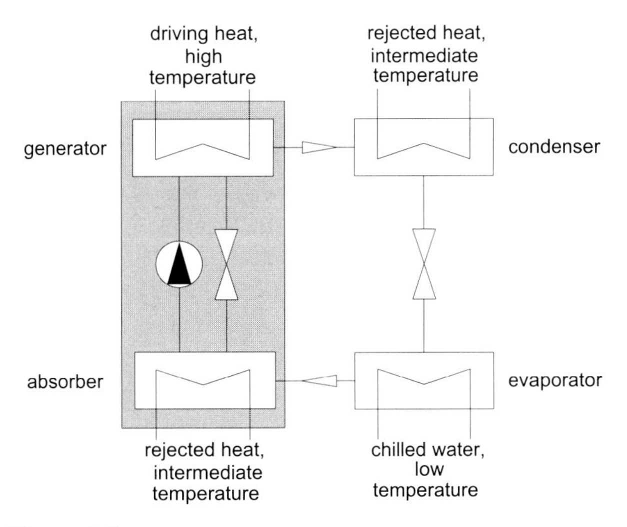

Figure 3.3
Schematic drawing of an absorption chiller for producing chilled water. The main energy input is the heat supplied to the generator. Electrical energy is necessary to drive the solution pump, unless a system with a bubble pump is used.

In the water/lithium bromide machine, water is the refrigerant, and cooling is based on the evaporation of water at very low pressures. Since water freezes below 0°C, the chilling temperature meets a physical limit at this level. LiBr is soluble in water if the LiBr mass fraction of the mixture is less than about 70 %. Crystallisation of the LiBr will occur at higher concentrations and may damage the machine. This sets a maximum temperature for the absorber. Poor control of temperature or a fast change of conditions may cause crystallisation. Appropriate operating controls will prevent this kind of problem. In order to sufficiently reduce the temperature of the absorber and dissipate the heat from the condenser, it is necessary to use a wet cooling tower. Water/lithium bromide chillers are mainly used for air-conditioning applications.

CYCLE PERFORMANCE

For solar-assisted air-conditioning systems with common solar collectors, single-effect absorption chillers are the most commonly used systems, because they require a relatively low temperature heat input. The term 'single-effect' refers to the fact that the supplied heat is used once by a single generator. Thermodynamic restrictions in the system dictate that the cooling capacity for ideal and real systems is always less than the heat input. The heat input to drive the generator, Q_{high}, must supply both the required heat of evaporation, r, that is required to vapourise the refrigerant out of the diluted water/lithium bromide solution and the heat of solution, l. However, the evaporator can only extract the heat of evaporation, r, that causes the refrigerant liquid to evaporate; the heat is extracted from the water containing the heat removed from the conditioned space, thus producing the cooling effect. Hence, the upper limit of the coefficient of performance for an ideal single-effect cycle of an absorption chiller is defined as follows:

$$COP_{thermal,\ max} = \frac{Q_{low}}{Q_{high}} = \frac{r}{r+l} \tag{3.3}$$

Typical COP's for large single-effect machines lie in the range of 0.7 to 0.8.

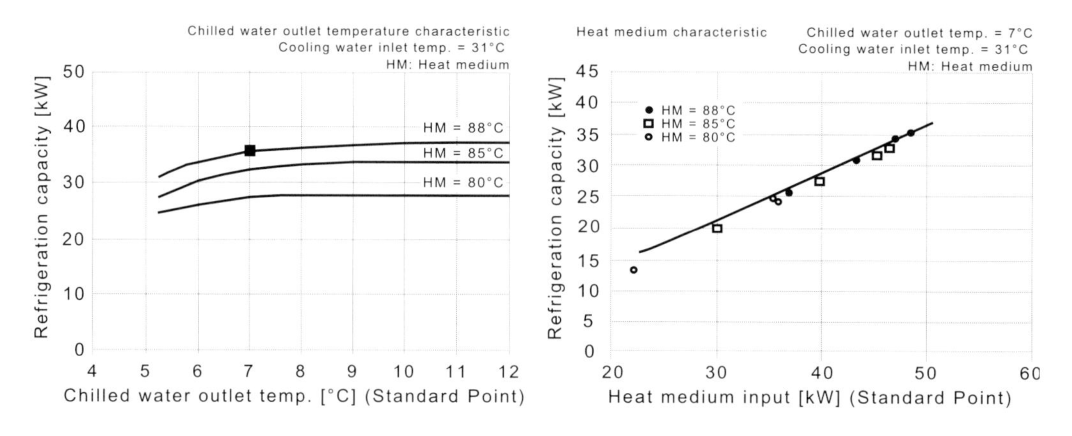

Figure 3.4
Manufacturer data of the absorption chiller WFC-SC 10 from the Yazaki company /3.2/.

Figure 3.5
Photograph of a 52 kW single-effect absorption chiller installed in a plant for solar air-conditioning of a wine cellar in Banyuls/France.

As an example, Figure 3.4 illustrates the performance characteristics of a commercially available single-effect absorption machine. The left part of the figure shows the refrigeration capacity versus temperature of chilled water outlet for different hot water inlet temperatures. The values are valid for a cooling water inlet temperature of 31°C. The figure indicates that it is possible to control the capacity of the chiller by varying the temperature of the hot water inlet. The right part of Figure 3.4 shows the refrigeration capacity as a function of the heat input. Most commercially available chillers also need an electricity supply to drive the solution pump, unless a bubble pump is used. In general, the internal electricity power consumption is in the range of 1 to 5 % of the chiller cooling power. The characteristic curves shown in Figure 3.4 refer to a chiller that employs an electric solution pump for transport of the diluted solution.

Double-Effect Absorption Chiller

A double-effect absorption chiller can be viewed as two single-effect cycles stacked on top of each other. The top cycle requires heat a higher temperature level compared to a single-effect machine. Generally, it is driven either directly by a natural gas or oil burner, or indirectly by supplying steam. Heat is added to the generator of the top cycle (primary generator), which generates refrigerant vapour at a higher temperature and pressure relative to the bottom cycle. The vapour is then condensed at this higher temperature and pressure, and the heat of condensation is used to drive the generator of the lower cycle (secondary generator), which is at a lower temperature. Double-effect cycles have a higher $COP_{thermal}$ than single-effect cycles. Typical operating COP's of double effect absorption chillers are close to 1.1 or slightly above and typical driving temperatures lie in the range of 140°C to 160°C. Current research is concentrating on three and four effect systems, which present an attractive potential for improved cooling performance, with a $COP_{thermal}$ of 1.7 to 2.2; but these systems require distinctly higher temperature of driving heat.

The need for higher driving temperatures makes double-effect chillers less suitable for solar-assisted air-conditioning systems using common solar collectors. It is possible to use high-efficient solar collectors to reach higher temperatures but this will increase the installation, operation and maintenance costs.

Manufacturer	Cooling power, type*	Driving T (°C)	Typical operation conditions, rated COP (if available)
Broad Air	20kW single and double effect	No data	No data
Colibri/Stork	100 kW NH_3 /H_2O single effect	> 90	$T_{cooling\ water}$ 27/32°C, $T_{chilled\ water}$ < 2°C: COP - 0.64
Coolingtec	70 kW R-134a/organic materials single effect	70-145	Example: T_{drive} 90°C, $T_{cooling\ water}$ 27°C, $T_{chilled\ water}$ 2°C: COP ~ 0.55
Dunham-Bush	327 kW single effect	Steam, 112	Chilled water 51 m^3/h, cooling water 105 m^3/h, steam 777 kg/h
EAW	15 kW single effect	75-95	Example: T_{drive} 85°C, $T_{cooling\ water}$ 30°C, $T_{chilled\ water}$ 12°C; COP - 0.70, hot/chilled water 2m³/h, cooling water 5 m³/h
Sanyo	105 kW single effect	85-95	Hot water 26.5 m^3/h, chilled water 8°C, cooling water 29.4°C
Trane	394 kW reversible single effect	> 100	No data
Yazaki	35 kW single effect	80-100	Chilled water 6 m^3/h, cooling water 14.5 m^3/h
York	420 kW double effect	> 116	Chilled water 65 m^3/h, cooling water 98 m^3/h

*All water/lithium-bromide unless otherwise indicated. Driving source: steam or hot water

Table 3.1
Examples of commercially available absorption chillers suitable for solar-assisted air-conditioning (only smallest available size included). The list does not claim to be exhaustive /3.3,/.

MARKET

Absorption chillers are commercially available from many manufacturers. The choice on the market is quite extensive, but most machines have a large capacity. Examples of commercially available absorption chillers suitable for solar-assisted air-conditioning are presented in Table 3.1. Only the smallest available size identified from each manufacturer is shown. An example of a 52 kW single-effect absorption chiller is shown in Figure 3.5. More information on this kind of technology can be found in literature (e.g., /3.4/).

3.1.2 Adsorption chillers

Instead of absorbing the refrigerant in an absorbing solution, it is also possible to adsorb the refrigerant on the internal surfaces of a highly porous solid. This process is called adsorption. Typical examples of working pairs are water/silica gel, water/zeolite, ammonia/activated carbon or methanol/activated carbon etc.. However, only machines using the water/silica gel working pair are currently available on the market. In absorption machines, the ability to circulate the absorbing fluid between the absorber and desorber results in a continuous loop. In adsorption machines, the solid sorbent has to be alternately cooled and heated to be able to adsorb and desorb the refrigerant. Operation is therefore intrinsically periodic with time. The cycle can be described as follows (see Figure 3.6):

Adsorption Cycle

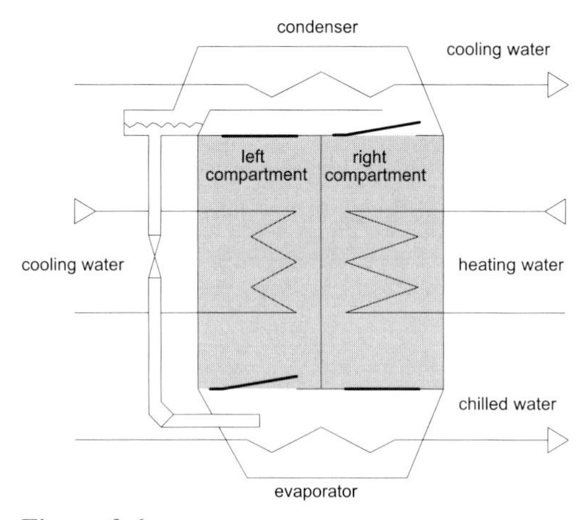

Figure 3.6
Schematic drawing of an adsorption chiller.

1. The refrigerant previously adsorbed in the one adsorber is driven off by the use of hot water (right compartment);

2. The refrigerant condenses in the condenser and the heat of condensation is removed by cooling water;

3. The condensate is sprayed in the evaporator, and evaporates under low pressure. This step produces the useful cooling effect;

4. The refrigerant vapour is adsorbed onto the other adsorber (left compartment). Heat is removed by the cooling water.

Cycle Performance

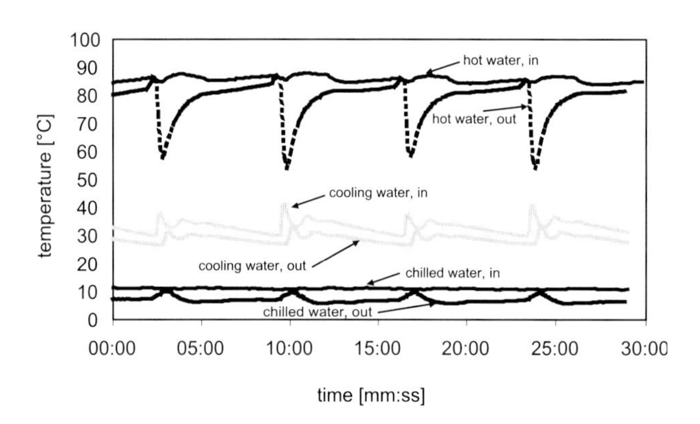

Figure 3.7
Typical return and supply temperatures of an adsorption chiller at the different temperature levels.

Once the right compartment has been fully charged and the left compartment fully regenerated, their functions are interchanged. In between, the two chambers may be directly coupled in order to achieve some heat recovery, since the hot chamber has to be cooled in the next step and vice versa. The time dependent temperatures in an adsorption chiller are shown in Figure 3.7; it can be seen, for example, that for this particular machine, a periodic change between the two compartments always takes place after about seven minutes.

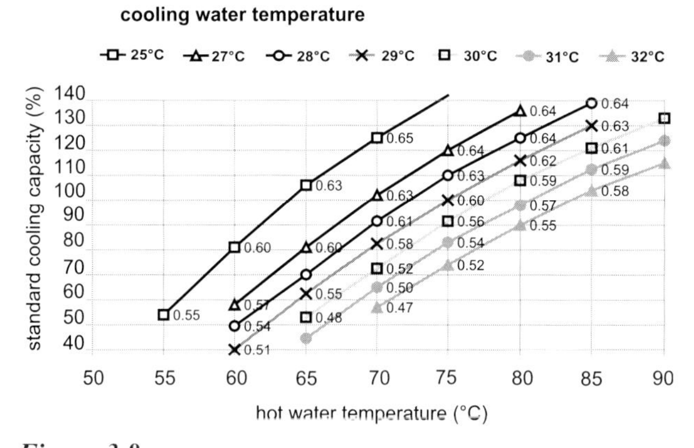

Figure 3.8
COP and cooling power of a commercially available adsorption chiller as a function of hot water temperature at different cooling water temperatures (chilled water 14/9°C) /3.5/.

In Figure 3.8, the cooling power and efficiency characteristics of a commercially available adsorption chiller are presented. The curves demonstrate that the machine can be run with rather low driving temperatures down to 55°C.

Market

Only a few manufacturers produce adsorption chillers. The performance characteristics of some commercially available adsorption chillers are summarised in Table 3.2. An example of a 70 kW adsorption chiller is shown in Figure 3.9.

Manufacturer	Cooling power	Driving temperature (°C)	Design conditions and rated COP
Mycom	70 kW water/silica gel	60 - 90	T_{drive} 80°C $T_{cooling\ water}$ 30°C $T_{chilled\ water}$ 9°C COP 0.61
Nishiyodo	105 kW water/silica gel	55 - 95	T_{drive} 80°C $T_{cooling\ water}$ 28°C $T_{chilled\ water}$ 8°C COP 0.61

Table 3.2
Commercially available adsorption chillers suitable for solar-assisted air-conditioning (only smallest available size included). This list does not claim to be exhaustive.

Figure 3.9
Adsorption chiller from the manufacturer Nishyodo; the machine is used for solar-assisted air-conditioning of a laboratory building at a hospital in Freiburg/Germany.

3.1.3 Vapour compression chillers

The most common refrigeration process applied in air-conditioning is the vapour compression cycle. Most of the cold production for air-conditioning of buildings is generated with this type of machine. The process employs a chemical refrigerant, e.g., R134a. A schematic drawing of the system is shown in Figure 3.10.

COMPRESSION CYCLE

In the evaporator, the refrigerant evaporates at a low temperature. The heat extracted from the external water supply is used to evaporate the refrigerant from the liquid to the gas phase. The external water is cooled down or - in other words - cooling power becomes available. The key component is the compressor, which compresses the refrigerant from a low pressure at low temperature to a higher pressure (high temperature) in the condenser.

Electrical energy is consumed by the motor used to drive the compressor. Thus, it is possible to reject the heat from the refrigerant at a higher temperature; for this purpose, either direct air cooling or a wet cooling tower is used. In the next step, the expansion valve throttles the pressure to the necessary pressure in the evaporator.

CYCLE PERFORMANCE

The coefficient of performance of a conventional vapour compression chiller, COP_{conv}, is defined as the ratio of the chiller's cold production, i.e., the rate of heat removal by the refrigerant, to the input power for the compressor.

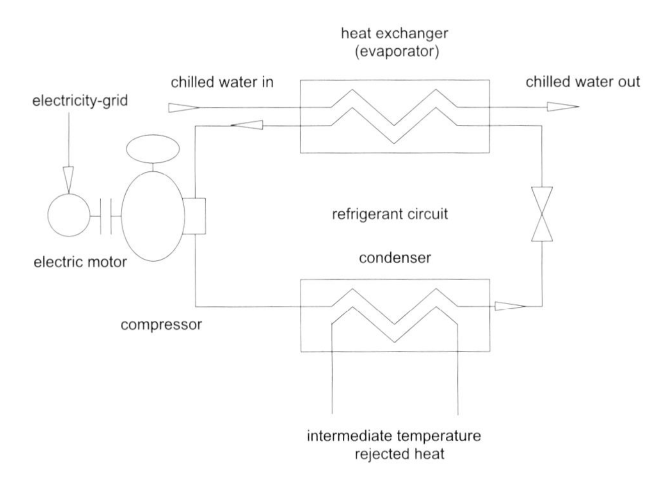

Figure 3.10
Schematic drawing of a vapour compression chiller.

Typical COP_{conv} values and capacity ranges of the most common compression machines are:

Reciprocating compressors:
COP_{conv} : 2.0 - 4.7;
Cooling capacity 10 - 500 kW

Screw compressors:
COP_{conv} : 2.0 - 7.0;
Cooling capacity 300 - 2000 kW

Centrifugal compressors:
COP_{conv} : 4.0 - 8.0;
Cooling capacity 300 - 30000 kW

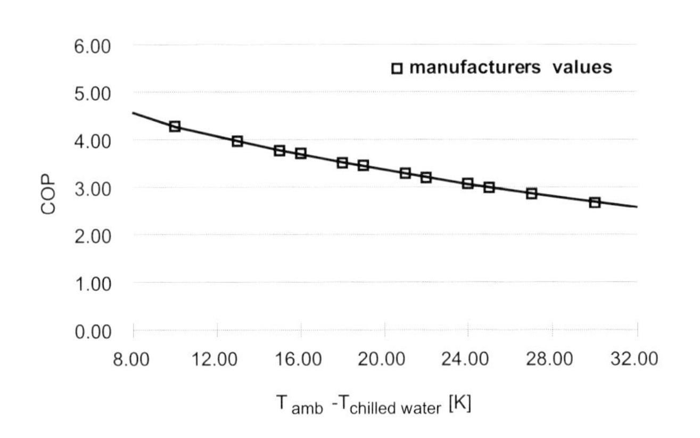

Figure 3.11
Performance of a commercial air cooled vapour compression chiller (R 134a, cooling power 150 kW) as a function of the temperature difference between ambient air, T_{amb}, and chilled water outlet, $T_{chilled\ water}$.

The COP of vapour compression chillers depends on the pressure (temperature) difference between evaporator and condenser. A higher pressure difference leads to a reduced COP. A typical curve for the COP_{conv} of an typical vapour compression chiller is shown in Figure 3.11. It displays the dependence of the chiller's performance on the temperature difference between evaporator and condenser. Concepts that make lower temperature differences possible are therefore beneficial since they reduce the energy consumption of the process.

3.2 Desiccant cooling systems

The use of sorption air dehumidification - whether with the help of solid desiccant material or liquid desiccants - opens new possibilities in air-conditioning technology. This can offer an alternative to classic compression refrigeration equipment. Alternatively, if it is combined with standard vapour compression technology, it leads to higher efficiency by an increase of the evaporator temperature of the compression cycle. Desiccant systems are used to produce conditioned fresh air directly. They are not intended to be used as systems where a cold liquid medium such as chilled water is used for heat removal, e.g., as for thermally driven chiller based systems. Therefore, they can be used only if the air-conditioning system includes some equipment to remove the surplus internal loads by supplying conditioned ventilation air to the building. This air-flow consists of ambient air, which needs to be cooled and dehumidified in order to meet the required supply air conditions. Desiccant cooling machines are designed to carry out these tasks.

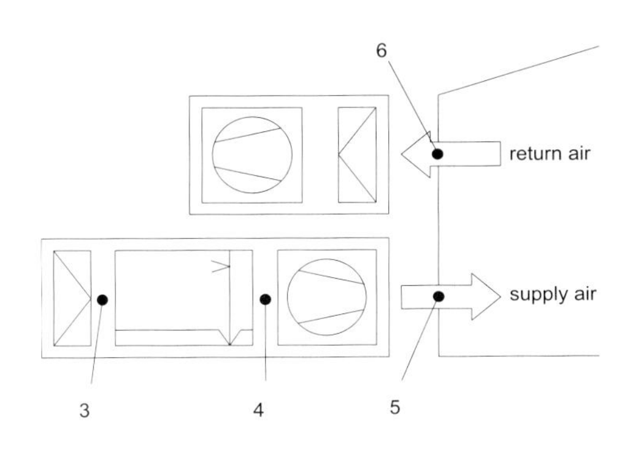

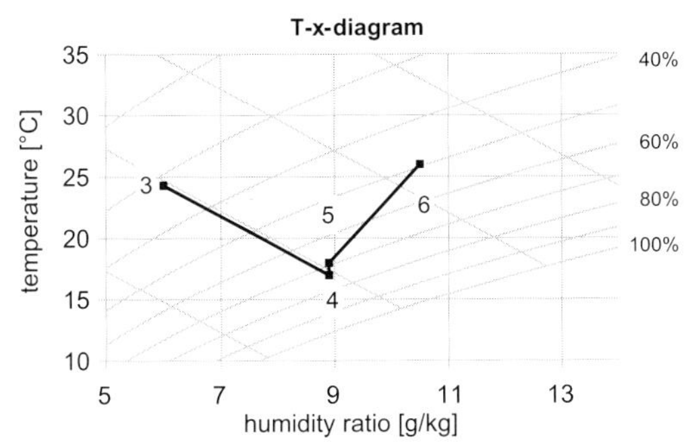

Figure 3.12
Schematic drawing of a direct evaporative cooling process.

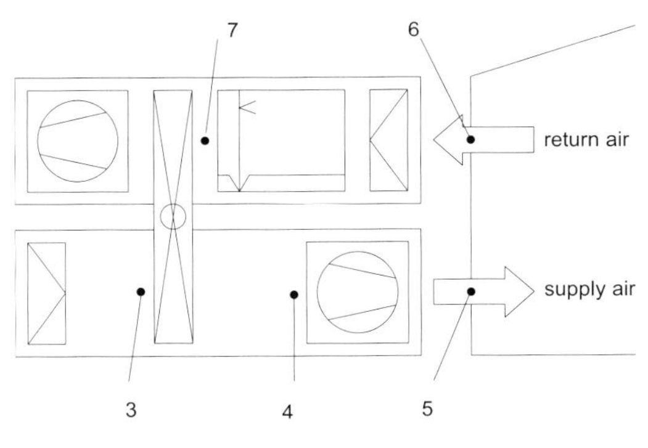

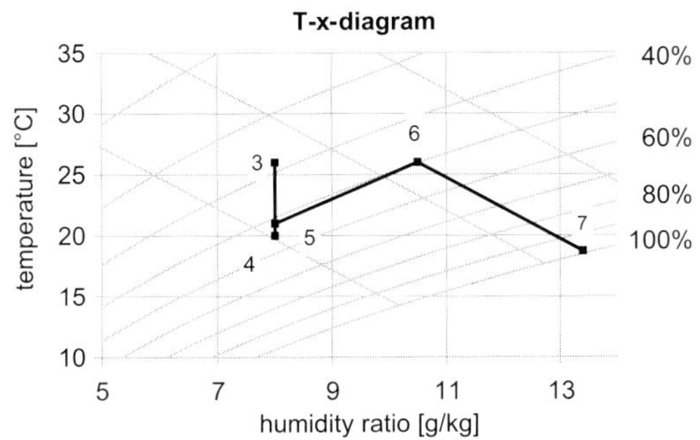

Figure 3.13
Schematic drawing of an indirect evaporative cooling process.

Economic advantages arise for desiccant cooling equipment when it is coupled with district heating or heat supplied from a combined heat and power (CHP) plant. Of particular interest is the coupling with thermal solar energy. The components of such systems are generally installed in an air-handling unit and are activated according to the operation mode of the air-conditioning system. These operation modes implement different physical processes for air treatment, depending on the load and the outdoor air conditions. These systems are based on the physical principle of evaporative and desiccant cooling. Unsaturated air is able to take up water until a state of equilibrium, namely saturation has been achieved. The lower the relative humidity of the air, the higher is the potential for evaporative cooling.

Evaporative Cooling Process

The evaporative cooling process uses the evaporation of liquid water to cool an air stream. The evaporation heat that is necessary to transform liquid water into vapour is partially taken from the air. When water comes into contact with a primary warm air stream it evaporates and absorbs heat from the air, thus reducing the air temperature; at the same time, the water vapour content of the air increases. In this case, the supply air is cooled directly by humidification and the process is referred to as direct evaporative cooling (Figure 3.12).

Indirect Evaporative Cooling Process

Indirect evaporative cooling involves the heat exchange with another air stream (usually the exhaust air), which has been previously humidified and thus cooled (Figure 3.13). In this case, the water vapour content of the primary air stream is not influenced.

Combined Evaporative Cooling Process

These two techniques of evaporative cooling can also be combined, in a process that is known as combined evaporative cooling (Figure 3.14).

Complementing combined evaporative cooling with desiccant dehumidification enhances the cooling capacity of the cycle and thus it is possible to reach even lower temperatures. This combined cooling process is referred to as desiccant cooling.

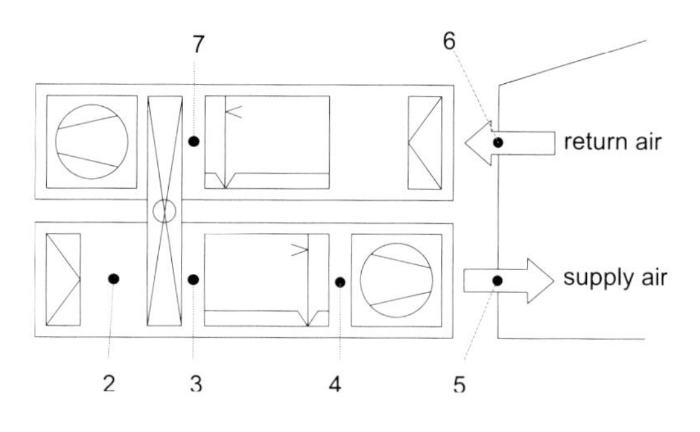

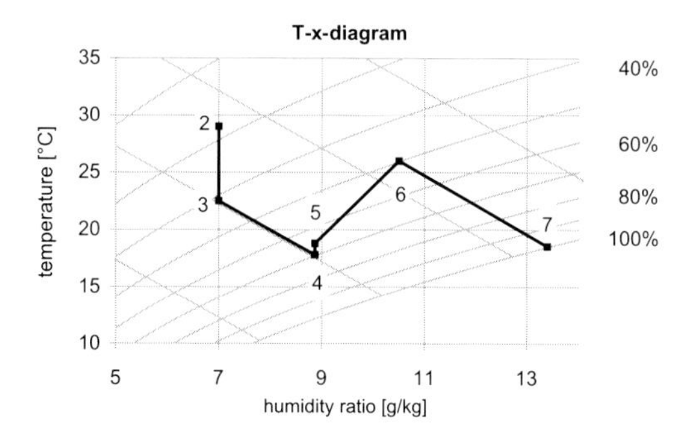

Figure 3.14
Schematic drawing of a combined evaporative cooling process.

Using evaporative cooling, either direct, indirect or in a combined process, it is not possible to reduce the vapour content of the ventilation air. But, using a desiccant cycle, in principle lowering of the temperature and the humidity ratio of ventilation air is possible.

Fresh air conditions have a considerable effect on the amount of cooling that can be achieved. If the outdoor air is properly pre-treated, the ventilation air can be cooled to lower temperatures via subsequent indirect and direct evaporative cooling. For this purpose, the pre-treatment involved is the desiccant dehumidification process to enhance the potential of evaporative cooling without obtaining a disproportionate high humidity ratio.

DESICCANT SYSTEM CONSTRUCTION

The dehumidification process uses either liquid or solid desiccants. Systems working with solid desiccant materials use either rotating wheels or periodically operated, fixed-bed systems. Systems employing liquid desiccants use air-desiccant contactors in the form of packed towers or the like.

Regeneration heat must be supplied in order to remove the adsorbed (absorbed) water from the desiccant material. The required heat is at a relatively low temperature, in the range of 50 to 100°C, depending on the desiccant material and the degree of dehumidification. Moreover, the solar desiccant cooling systems, depending on the cooling loads and environmental conditions, will use one of the abovementioned cooling modes, i.e,. direct evaporative cooling and/or indirect evaporative cooling and/or desiccant cooling, with the aim of providing comfort conditions in the building.

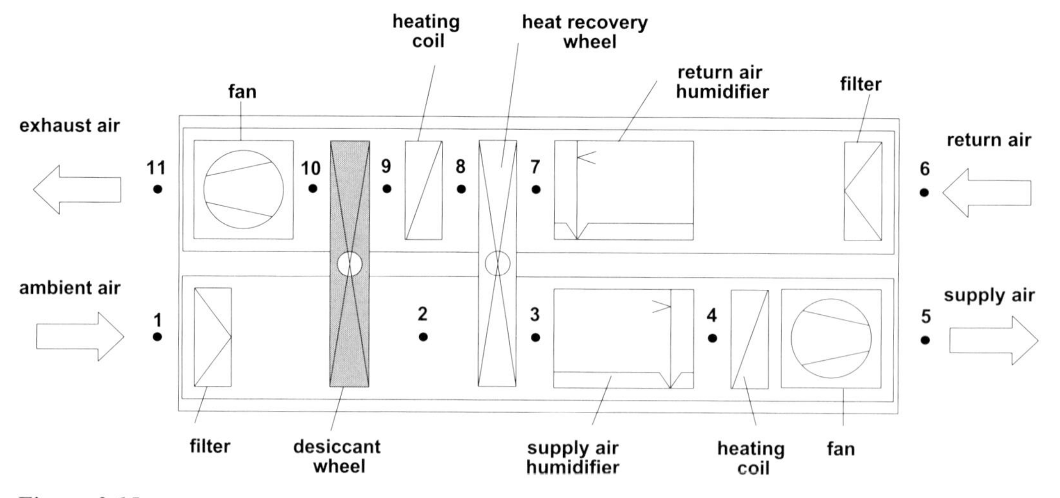

Figure 3.15
Schematic drawing of a desiccant cooling air-handling unit.

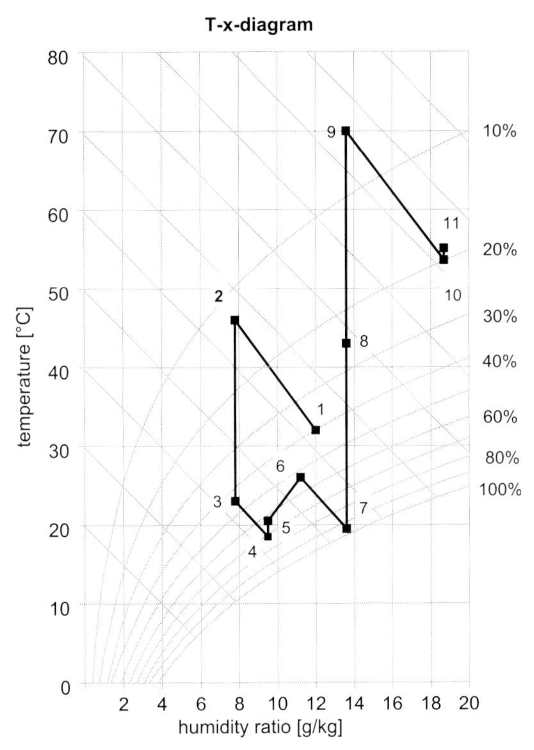

Figure 3.16
Typical desiccant cooling process in the T-x-diagram.

The most commonly used desiccant cooling process, which is based on the use of desiccant wheels, works as follows (see Figure 3.15 and 3.16):

PROCESS DESCRIPTION - SUPPLY AIR

The ambient air (1) is dehumidified in a desiccant wheel, causing the air temperature to increase; the process is nearly adiabatic (2). The regenerative heat recovery leads to cooling of the air inlet to the humidifier, by means of indirect evaporative cooling (3). Depending on the air inlet temperature and humidity supplied, the temperature is reduced by direct evaporative cooling in the humidifier, with a simultaneous increase in humidity up to condition (4). The coil on the supply stream is in operation only for heating conditions. The fan releases heat, leading to an increase in the temperature of the supply air, which brings about the supply air condition (5). An increase in temperature of up to 1°C is usually expected. A proper design of the fan is recommended so as to minimise the heat added to the supply air.

PROCESS DESCRIPTION - RETURN AIR

The return air from the room is in state (6). The air is then humidified as close as possible to saturation (7). This state is the one which guarantees the maximum potential for indirect cooling of the supply air stream through the heat exchanger for heat recovery. The heat recovery from (7) to (8) leads to an increase in the temperature of the air, which is then used as regeneration air. The air is subsequently reheated in the coil until it reaches state (9). The temperature of the latter is adjusted such as to guarantee the regeneration of the sorption wheel (9 to 10).

Figure 3.17
Example of a desiccant air-handling unit with desiccant wheel (nominal air-flow 4500 m³/h).

It is important to mention that in many desiccant systems a bypass is installed which allows that some of the air coming from the heat recovery unit bypasses the regeneration air heater and the desiccant wheel. Depending on the actual conditions up to more than 20 % of the air can go through the bypass thus saving regeneration heat and also electricity because of the reduced pressure drop along the desiccant wheel.

The $COP_{thermal}$ of a desiccant cooling system is defined as the ratio between the enthalpy change from ambient air to supply air, multiplied by the mass air-flow, and the external heat delivered to the regeneration heater, $\dot{Q}_{reg}$:

$$COP_{thermal} = \frac{\dot{m}_{supply}\ (h_{amb} - h_{supply})}{\dot{Q}_{reg}} = \frac{\dot{m}_{supply}\ (h_1 - h_5)}{\dot{Q}_{reg}} \tag{3.4}$$

Operation Mode

The value of $COP_{thermal}$ of a desiccant cooling system depends strongly on the conditions of ambient air, supply air and return air. Under normal design conditions, a $COP_{thermal}$ of about 0.7 is achieved and the cooling power lies in the range of about 5-6 kW per 1000 m^3/h of supply air.

Average prices for an entire desiccant air-handling unit using a sorption wheel for dehumidification range from 5 to 10 € per m^3/h of nominal supply air-flow. The price strongly depends on the nominal air-flow rate of the system which is directly proportional to the cross-sectional area of the wheel.

An example of a desiccant air-handling unit with a configuration such as in Figure 3.15 is shown in Figure 3.17. The key components of desiccant cooling systems are described in the following sub-chapters.

3.2.1 Desiccant wheel

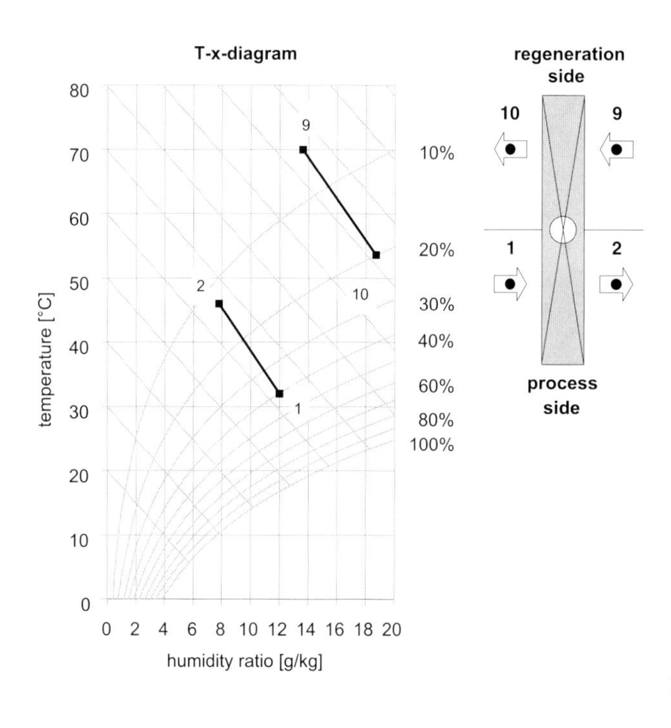

Figure 3.18
Psychometric chart for moist air showing the state changes for dehumidification of air in a desiccant wheel.

The rotary dehumidifier, known as a desiccant wheel, is a common type of sorption dehumidifier based on the use of a solid desiccant. The desiccant material is coated, impregnated or formed in place on a supporting rotor structure, similar to that of a common rotary heat exchanger. Typically, the basic material which forms the supporting structure is a mix of different fibres, including glass, ceramic binders and heat-resistant plastics. A desiccant wheel functions as a heat and mass exchanger between two air streams, i.e., supply and return, and can be operated either as a dehumidifier or as an enthalpy recovery component in desiccant cooling and heating, respectively.

Depending on the chosen operating mode (dehumidifying or enthalpy recovery), the rotational speed of the wheel varies commonly within the ranges of 6 - 12 rotations per hour for the dehumidifier mode and 8 - 14 rotations per minute for the enthalpy recovery mode, respectively. The physical process taking place across the wheel is further described only for the dehumidifier mode; the process for the enthalpy recovery mode is similar to that of usual enthalpy recovery wheels.

Process description

Figure 3.18 is a schematic illustration of the state changes occurring during the dehumidification of air in a desiccant wheel in a psychometric chart. The numbers refer to those in Figure 3.16. The supply air is dehumidified (state change 1-2) on the process side of the rotor. The return air, after being heated, flows through the regeneration side of the rotor (state change 9-10) causing desorption of the water that was bound in the desiccant on the process side. For a given desiccant wheel operating under conditions of given air-flow rates and a given speed of rotation, the state of the out

Figure 3.19
Example of a sorption desiccant wheel integrated into a cassette (Source: Klingenburg GmbH, Germany).

going process air (2) depends primarily on the incoming air states 1 and 9.

REGENERATION

The desiccant is regenerated/reactivated by supplying heat to the regeneration air, which is sufficient to raise the temperature of the desiccant to a value at which the vapour pressure of the water bound in it exceeds the partial pressure of the water vapour in the warm regeneration air. The change in the state of the air on the process side occurs with an increase in enthalpy. The energy associated with the sorption and desorption processes is equal to the latent heat of condensation plus a differential heat of sorption. It is beneficial to have a low total heat of sorption. In addition, the state change is also affected by the heat stored in the rotor matrix on the regeneration side.

As it was mentioned above, the two previous operation modes are a function of the rotational speed of the wheel. At a high rotational speed, the activated sorption capacity of the wheel is reduced. The process then can be represented on the psychrometric chart along a line which connects both inlet air points for the two streams; this corresponds to an enthalpy recovery behaviour which is favourable for winter applications, i.e., heat and moisture recovery from the return stream.

DESICCANT WHEEL MANUFACTURES

Table 3.3 gives a list of desiccant wheel manufacturers, along with a short description of the available products. It is difficult to provide market prices for sorption regenerators since they are not an independent market product but are purchased by manufacturers of air-handling units. Also the actual price depends strongly on the size of the air-handling unit.

Table 3.3
Manufacturers and product description of sorption dehumidifiers /3.6/. The list does not claim to be exhaustive.

Company	Country of Origin	Desiccant	Wheel Size
Munters USA	US	SiGel, AlTi, Silicates, New Proprietary	0.25 - 4.5 m
Munters AB	Sweden	SiGel, AlTi, Silicates, New Proprietary	0.25 - 4.5 m
Seibu Giken	Japan	SiGel, Am, Silicates, New Proprietary	0.1 - 6 m
Nichias	Japan	SiGel, Mol. Sieves	0.1 - 4 m
DRI	India	SiGel, Mol. Sieves	0.3 - 4 m
Klingenburg	Germany	Al oxide, LiCl	0.6 - 5 m
ProFlute	Sweden	SiGel, Mol. Sieves	0.5 - 3 m
Rotor Source	US	SiGel, Mol. Sieves	0.5 - 3 m
NovelAire	US	SiGel, Mol. Sieves	0.5 - 3 m

3.2.2 Humidifiers

Humidifiers are common components in air-handling units. In desiccant cooling systems, their main task is - at least during the cooling season - to cool the air by means of evaporative cooling rather than to increase the air humidity ratio.

PRINCIPLE

Air humidification can take place in two thermodynamically different ways:

1. Injection of water vapour: in this case, an increase in mixed temperature which is dependent on the steam condition (wet steam, saturated steam, superheated steam) occurs; however, this can be frequently ignored for practical calculations (quasi-isothermal change in condition).

2. Injection of liquid water: if small water/air fractions, e.g., smaller than 0.3 kg water per m^3 of air, are realised, almost an adiabatic change of condition takes place which is always accompanied by mixed cooling with a main contribution of evaporation cooling.

Construction Types

Obviously, only the latter air humidification systems and equipment are relevant for desiccant cooling processes; so only these are considered further. There are various direct humidification processes and systems including:

- Spray nozzle humidifier (air washer)
- Two-component alloy humidifier
- High-pressure nozzle sprayer (cold steam generator)
- Mechanical motor-driven sprayer
- Ultrasonic humidifier
- Evaporation humidifier (contact or trickling humidifier)
- Hybrid humidifier

An overview of the technical construction of the most common humidifier types which can be used in desiccant cooling systems is given in Table 3.4

Humidifier type	Contact humidifier	Spray nozzle humidifier	High-pressure nozzle sprayer	Ultrasonic humidifier	Hybrid humidifier
General information					
Humidification process	Airflow along wetted elements	Airflow through water curtain of fine water droplets from spray nozzles	Compressed air (sometimes more than 70 bar) atomises the water in opposite direction to the airstream	Ultrasonic vibrations cause the airstream to accept a spray of very fine water droplets directly	Atomising water under medium pressure (4-8 bar). then cold evaporisation on porous ceramic V-material
Max. humidification efficiency [%]	65 - 90	100	95 - 100	75 - 95	85 - 90
Water circuit					
Water treatment	Not necessary. softening recommended	If demineralised-water, then adapted materials required	Demineralised-water from reverse osmosis	Demineralised-water from reverse osmosis	Reverse osmosis, addition of silver ions
Water pressure [bar]	1 - 10	up to 3	30 - 180	1- 10	7
Recirculation[1] [%]	50	15	0	0	15
Excess water / drain quantity[1]	20	15	0	5	15
Air circuit					
Length of air humidification section [mm]	600 - 1000	1500 - 3000	1500 - 2500	2000 - 3000	1200 - 1500
Air pressure drop [Pa]	90 - 250	70 - 200	50	up to 20	70
Oversaturation	No	Possible	Yes	Yes	No
Control (on/off or modulating)					
Method	Dewpoint control; controllable bypass on air side	Water supply	Water supply	Amplitude of vibration	7 - stage water supply control on nozzles + water
Controllability	moderately	good	moderately	good	good
Maintenance					
Degree of maintenance	medium	high	high	moderate	moderate
Electric consumption					
Electric power in W per kg/h	5	10	6 - 15	1 - 5	-[2]

[1] The percentage of recirculation and drainage relates to the quantity of water supplied to the air
[2] no information available

Table 3.4
Different humidifier types and their specifications.

3.2.3 Air-to-air heat exchangers

The application of air-to-air heat exchangers is essential for the operation of a desiccant cooling system. Following the general process description in a psychometrics chart according to Figure 3.16, the function of this heat exchanger is to shift the supply air from a state with high temperature to a state with low temperature. This is carried out by decreasing the temperature of the supply air in counter-flow to the return air, by means of a sensible heat exchange, i.e., without any change in humidity ratio.

IMPORTANCE OF HIGH EFFICIENCY

The influence of the effectiveness of the air-to-air heat exchanger on the performance of the desiccant cooling cycle in summer is demonstrated in Figure 3.20. The effectiveness, ε, of the heat exchanger is defined as follows:

$$\varepsilon = \frac{\Delta T_{realised}}{\Delta T_{max}} \quad (3.5)$$

where ΔT_{max} is the maximum possible temperature difference between the hot inlet of the supply air and the cold inlet of the return air,

$$\Delta T_{max} = T_{supply,\ in} - T_{return,\ in} \quad (3.6)$$

and $\Delta T_{realised}$ is the temperature difference of the supply air between the inlet and outlet of the heat recovery unit:

$$\Delta T_{realised} = T_{supply,\ in} - T_{supply,\ out} \quad (3.7)$$

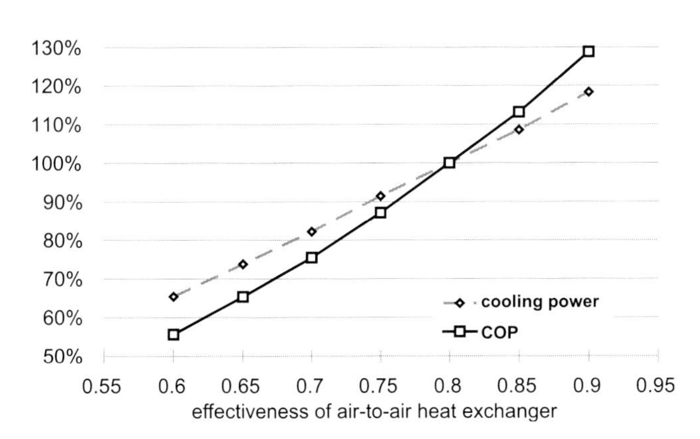

Figure 3.20
Influence of the air-to-air heat exchanger effectiveness on the performance of a standard desiccant cooling cycle (values for an efficiency of 0.8 were set to 100 %).

In Figure 3.20 the reference value of the effectiveness ε was set to 0.8. For this value the cooling capacity of the desiccant cooling cycle as well as the $COP_{thermal}$ is set to 100 %. Reducing the effectiveness to a value of 0.6 reduces the cooling power by nearly 35 % and increasing the effectiveness to 0.9 increases it by about 18 %. At the same time the $COP_{thermal}$ of the cycle is reduced by nearly 45 % for an effectiveness of 0.6 and is increased by more than 28 % for an effectiveness of 0.9, respectively. This underlines the necessity to install a highly efficient air-to-air heat exchanger in desiccant cooling units.

CONSTRUCTION TYPES

The typical application of highly efficient air-to-air heat exchangers is heat recovery of the ventilation air in mechanical ventilation systems during the heating season. The technical solutions to achieve highly efficient heat recovery and their main specifications are summarised in Table 3.5. While rotor-based heat recovery units in general achieve high effectiveness, this is not generally valid for the other types. Air-to-air plate heat exchangers and air-water-air circuits in general achieve effectiveness values in the range of 0.4 to 0.6. In order to apply them in desiccant cooling systems they have to be designed carefully, employing highly efficient components with a domination of the counter-flow regime.

type	wheel	plate	air-water-air circuit
description	a wheel typically consisting of an aluminium matrix rotates from the supply air to the return air (typically 8 - 10 rotations per minute)	plate heat exchanger typically constructed of plastic materials or aluminum	two air-to-water heat exchangers are connected by a water circuit; a water pump circulates the water, which transfers heat from one air stream to the other
volume	medium	large, if highly effective	large, if highly effective
effectiveness	typically high (around 0.8)	typically 0.4 - 0.6 up to 0.85 possible; requires specific construction (counter-flow design)	typically 0.4 - 0.5 high value requires specific construction (counter-flow design of each water-air heat exchanger)
price	average	average	high
advantages	· standard component in ventilation heat recovery systems · highly effective · compatible with desiccant rotor	· airstreams physically separated · possible to combine with highly effective indirect evaporative cooling · low maintenance	· air streams physically separated · supply and return air need not be combined in the same air-handling unit
possible application in desiccant cooling	standard type of air-to-air heat exchanger in desiccant cooling	application in systems with return and supply air physically separated (requires that regeneration is not achieved with return air but e.g., with ambient air)	application in systems with return and supply air physically separated and not combined in the same air-handling unit (requires that regeneration is not achieved with return air but e.g., with ambient air)

Table 3.5
Specifications of air-to-air heat exchanging equipment.

3.2.4 Desiccant cooling with liquid sorbent

Liquid sorbent agents can also be used for the dehumidification of conditioned air. The liquid desiccant system is essentially an open-cycle absorption system, where water serves as the refrigerant. However, whereas a large number of working fluid pairs are available for closed absorption refrigerating machines, there is only a small number of suitable materials for open liquid-based systems which can be used for the conditioning of ventilation air. This is due to the strict limitations that apply to aqueous, hygroscopic solutions, since they come in direct contact with the environment. The solutions used should be non-toxic and environmentally friendly, and should not contain any volatile material other than water. In practice, liquid sorbent agents which consist principally of salts dissolved in water are mainly used, e.g., lithium chloride or calcium chloride. These hygroscopic salts lower the vapour pressure of water in solution sufficiently to absorb humidity from the air. In contrast to the case of the solid sorbents, the water bonding mechanism is not adsorption, but absorption.

Main Components

The sorption systems used for the drying of air consist basically of an absorber and a regenerator, as shown in Figure 3.21. These are air-solution heat and mass exchangers, normally in the form of packed towers, where air and solution come into contact in counter-flow or cross-flow. Both types of equipment may be identical in structure, i.e., they have the same type of exchange surfaces and usually differ only in terms of their relative dimensions. Humidity is absorbed from the process air into the hygroscopic solution in the absorber. Then the salt solution is regenerated so that the same initial concentration is always available when drying the air. In order to remove the absorbed water out of the dilute solution, heat at a relatively low temperature level is required; temperatures of about 60°C to 70°C are sufficient.

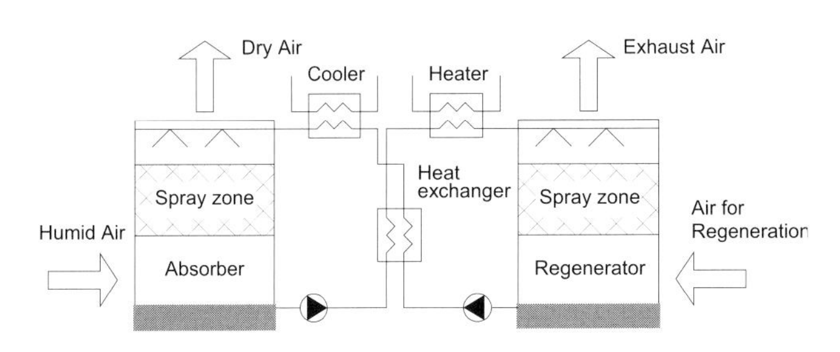

Figure 3.21
Schematic drawing of a liquid sorption system.

HEAT RECOVERY

Referring to Figure 3.21, it is interesting to note the solution heat exchanger, which recovers heat from the hot desiccant coming from the generator on its way to the absorber, and uses it to preheat the cold desiccant leaving the absorber on its way to being regenerated. Such internal heat recovery, typical for closed absorption systems, is more difficult to realise with solid desiccant systems where the desiccant does not flow.

ENERGY STORAGE

Another advantage of liquid desiccant systems is the ability to store cooling capacity by means of the regenerated desiccant. Thus, hygroscopic salt solution may be concentrated when solar energy is available, and used to dehumidify process air later, when needed. The dehumidification process can be operated as long as a concentrated desiccant is available and is independent of the availability of driving heat for regeneration at the same time. This form of cold storage is the most compact, requires no insulation and can be applied for indefinitely long periods of time.

HIGH CONCENTRATION DIFFERENCES

The concentration difference between concentrated and diluted solution can be increased by cooling the absorption process and using a special design of the absorber. To cool the absorption process, either an air-cooled or water-cooled absorber may be employed. This feature of high concentration difference increases the use of energy storage by separating concentrated and diluted desiccant.

MARKET SITUATION

Only a few manufacturers offer liquid sorption systems at present. However, the commercially available systems are not well adapted to the use of solar energy for desiccant regeneration. Pilot plants of solar-driven systems are in operation in several demonstration projects.

3.3 Other components of air-conditioning systems

Some further components are important parts of air-conditioning systems and shall be described here briefly, mainly focussing on their role concerning the energy performance of a solar-assisted air-conditioning system.

3.3.1 Cooling towers

SYSTEM TYPES

A cooling tower is a specialised heat exchanger where cooling water is brought into contact with ambient air to transfer rejected heat from the coolant to the ambient. For this purpose, two basic types of systems can be found: open-circuit systems, where there is direct contact between the primary cooling-water circuit and the air, and closed-circuit systems where there is only indirect contact between the two fluids across heat exchanger walls. Open-circuit systems are commonly known as ‘open cooling towers’, ‘wet cooling towers’ or just as ‘cooling towers’. A characteristic feature of all such systems is, that they mostly use latent heat transfer where the coolant, which has to be water, is cooled by evaporating about 2-3 % of the coolant itself.

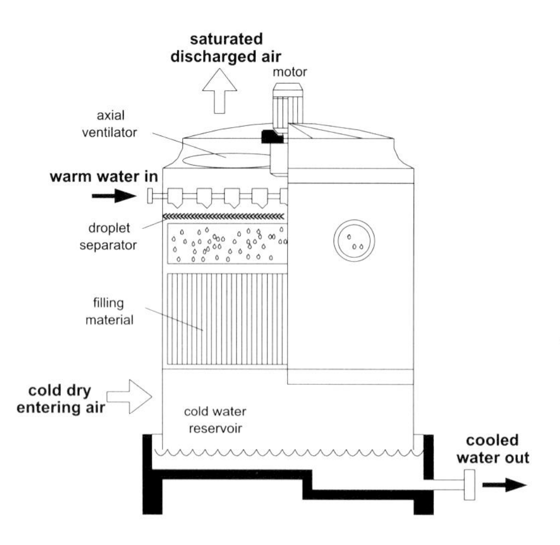

Figure 3.22
Schematic drawing of an open type wet cooling tower.

This results in highly efficient cooling operation, even at coolant temperatures below ambient temperature, together with minimum investment cost, but it is accompanied by significant water consumption at any operational state. Closed-circuit systems, on the other hand, show a great variety of types and operational modes. These range from dry air coolers, transferring just sensible heat to ambient air, to a second type of wet cooling tower, incorporating an auxiliary water circuit for spraying heat exchanger tube bundles at the air side and primarily utilising latent cooling. In addition there are several hybrid systems which combine both cooling modes, latent cooling by evaporation of water and sensible cooling against ambient air, or which are able to switch between both cooling modes in dependence on the operation.

However, all closed-circuit systems generally show less efficient operation, increased electricity consumption due to larger fans and at least doubled investment costs in comparison to open-circuit cooling towers. Further here, only open-circuit cooling towers will be discussed. A schematic drawing of such a cooling tower is shown in Figure 3.22.

Operation Principle of Wet Cooling Towers

The basic function of a cooling tower is to ensure a good heat and mass transfer between the cooling water stream and ambient air. Thus the hot water enters the upper part of the cooling tower, where it is evenly distributed across the tower by a spraying system. To increase the effective contact surface between water and air, there is additional filling material installed inside the cooling tower. At the bottom of the tower, the cooled water is collected again in a reservoir. To ensure sufficient air-flow through the tower, a fan is installed that either forces entering air into the tower or sucks discharge air at the outlet. Both cross-flow and counter-flow designs are common and available, but a counter-flow arrangement with air entering at the bottom of the tower and a suction fan located at the top is the standard configuration in Europe. Additional installations for water treatment and blow-down are required for all cooling towers to replace the evaporated cooling water and to prevent fouling.

Performance and Selection Guidelines

The performance of a wet cooling tower mainly depends on the wet bulb temperature of the ambient air, while it is only slightly affected by ambient temperature. The design limit for the temperature of the cooling water leaving the tower is about only 3 - 5°C above the wet bulb temperature, which typically is still below ambient air temperature. During part load operation the reduction ratio of the water flowrate must not exceed 1 : 5 to avoid clogging, whereas it is even possible to switch the fan off when running at about 10 % cooling load. As a cooling tower operates with about 90 % latent cooling even at low ambient temperatures, the water evaporation can directly be estimated from the cooling load; however at least 50 % additional blow down has to be considered to obtain the total water consumption. Since there is a highly non-linear relation for air between temperature and water vapour saturation pressure, no simple equations can be given to describe the operational behaviour of a cooling tower at different operational states.

Design and Performance Figures

Typical design and performance figures for an open-circuit wet cooling tower are:

Air volume flow: 130 - 170 m^3/h per kW of cooling power.

Electricity consumption of the fan:	6 - 10 W per kW of cooling power for axial ventilators; 10 - 20 W per kW of cooling power for radial ventilators.
Control:	In order to save energy, it is recommended to equip the ventilator with a frequency control, so that the fan velocity can be adapted to the required cooling power.

3.3.2 Fans, pumps and accessories

Fans and pumps are conventional components of heating and cooling systems in buildings. Therefore there is no need to describe them in detail here. However, since they contribute to the overall energy consumption of a system, their energy performance is important for the overall efficiency. Major energy-related aspects are discussed below.

Fans

The main electricity consumption of a ventilation system is caused by the transport of air, which is achieved with fans. Therefore, it is important to consider the electricity consumption of the fans in the overall energy balance of a ventilation system. This is particularly valid for desiccant cooling systems, since additional components such as the sorption regenerator create an additional pressure drop, which contributes to an increased electricity consumption for air transport.

ENERGY PERFORMANCE

The minimum required energy flow for transport of air is given by the product of air volume flow and pressure drop along a component. For a complete ventilation system the total minimum power for air transport, P_{min}, is given by the sum of the respective value of all components:

$$P_{min} = \sum_i \Delta p_i \dot{V}_i \tag{3.8}$$

where i goes from 1 to n and n is the number of components (internal, i.e., inside the air-handling unit and external, i.e., in the duct system).

The air volume flow can vary along the system either due to different hydraulic cycles which are connected with diverters and mixers or due to temperature changes along the air-handling unit which lead to a related density change and thus to different air volume flows. The total electric power necessary to drive a fan is given by the minimum required power, P_{min}, multiplied by the efficiencies of the ventilator itself and of the electric motor, η_{vent} and η_{el}, respectively:

$$P_{el} = \frac{P_{min}}{\eta_{vent}\,\eta_{el}} = \frac{P_{min}}{\eta_{fan}} \tag{3.9}$$

For turbulent flow, the pressure drop along a system is nearly proportional to the square of the velocity and so also to the square of the air volume flow (exponents between 1.6 and 2 are found in practice, depending on the roughness of the duct system):

$$\Delta p \approx \dot{V}^2 \tag{3.10}$$

As a result, the required minimum power for air transport through a component is proportional to the cube of the air volume flow:

$$P_{min} = \Delta p\, \dot{V} \approx \dot{V}^3 \tag{3.11}$$

This expression underlines the necessity to control the air volume flow in systems where no constant volume flow is required, in order to reduce the electricity consumption. The best control is achieved by controlling the rotation speed of the ventilator, for instance with frequency controllers. Other means such as damper are less efficient, as can be seen from Figure 3.23.

CONSTRUCTION TYPES

A clear distinction can be made between axial and radial ventilators. Both types are used in air-handling units. Radial ventilators are available for very different pressure ranges:

- Low pressure: up to 700 Pa total pressure difference
- Medium pressure: 700 - 3000 Pa total pressure difference
- High pressure: 3000 - 30000 Pa total pressure difference

Axial ventilators are generally applicable in a lower pressure range. The main types are:

- Low pressure: up to 300 Pa total pressure difference
- Medium pressure: 300 - 1000 Pa total pressure difference
- High pressure: above 1000 Pa total pressure difference

Typical efficiency values, η_{vent}, for highly efficient radial ventilators are commonly higher than 85 % and for axial ventilators they are in the range of 65 - 85 %. The overall efficiency value of highly efficient fans, η_{fan}, at the design point is in the range of 60 - 75 %; this includes the efficiency of the electrically driven motor.

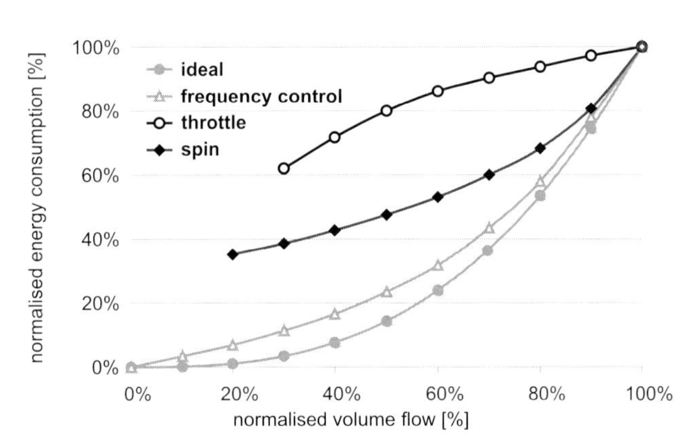

Figure 3.23
Normalised energy consumption for different ways to control the air volume flow in an air-handling unit. The ideal curve results from the minimum required energy for air transport in a system with a quadratic dependence of pressure drop on the air volume flow.

Pumps

Solar thermal systems require more complex hydraulic schemes than conventional heating or cooling systems. In general two-loop collector systems are used, at least in the case of large collector areas. Therefore, in most cases two additional pumps are required for the primary and secondary solar loops. In the primary loop either water or an anti-freezing fluid is used, depending on the climatic conditions at the installation site. This means that the pump in the primary solar loop has to be designed for an anti-freezing fluid as well as for high temperatures, since in a collector system which is designed for solar cooling, temperatures well above 100°C can be achieved, at least under stagnation conditions, i.e., when no heat is extracted from the collector loop and solar radiation is available.

ENERGY PERFORMANCE

In principle, the formulae given above for fans are valid in a similar way for pumps. Typical fluid-mechanics efficiency values of the rotor lie in the range of 40 - 60 % for small pumps, 60 - 75 % for medium-sized pumps and up to 85 % for large pumps, e.g., for those installed in large, central, chilled-water networks. The overall efficiency, including losses due to the electric motor ranges between 30 - 55 % for small pumps, 45 - 65 % for medium size pumps and 60 - 75 % for large pumps. As for fans, it is recommended to install frequency-controlled motors, at least in hydraulic cycles in which large variations of the required mass flow occur. Use of a variable-flow pump in a solar collector cycle allows - within certain limits - the collector outlet temperature to be controlled even under conditions of variable radiation on the collector surface (so called matched-flow concept).

Air filters and noise attenuators

Air filtration is one of the tasks of air-conditioning systems. Filters are commonly one of the parts of an air-handling unit. The performance of these components is specified by parameters describing the minimum dimensions and the number of contaminant particles able to bypass the filter. On the other hand, a filter also represents a non-negligible cause of resistance to the air motion. The induced pressure drop must be accounted for in fan dimensioning and thus electricity consumption assessment.

CLASSIFICATION

Air filters are divided into three main classes according to the materials of which the filters are made and the corresponding different physical effects occurring in the separation of particles:

- Filters for coarse dust.
- Filters for fine dust.
- Absolute filters, divided into:
  a) HEPA (High-Efficiency Particulate Air) filters
  b) ULPA (Ultra-Low Penetration Air) filters

Only coarse dust and fine dust filters are used for conventional air-conditioning purposes. In civil engineering, only in special applications such as hospitals, clean rooms or data-processing centres higher efficiency filters (HEPA, ULPA) are required.

The filter capacity depends on the set air speed through the filtrating material. Filters for fine dust and aerosol filters maintain their efficiency level without any significant changes even at greater variations from the rated load. This makes these filters very suitable in variable air volume (VAV) systems.

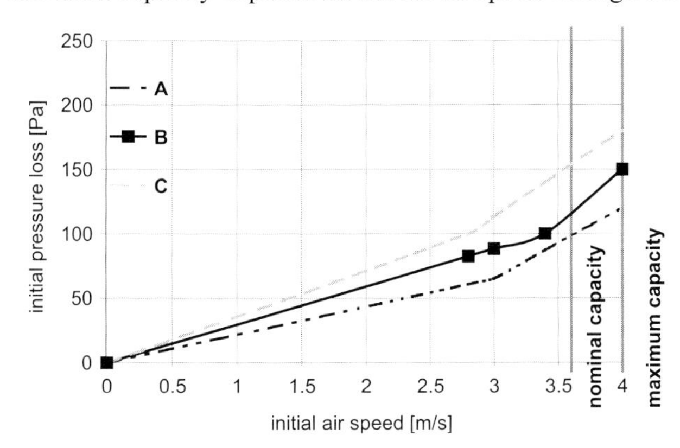

Figure 3.24
Pressure drop changes versus air speed for three models of high-efficiency bag filters.

PRESSURE DROP

The pressure drop caused by the filter is clearly related to air speed and to dust accumulation on the filter surface, which increases with the age of the filter. Figure 3.24 shows, as an example, how the pressure drop changes according to air speed for three models of high-efficiency bag filters.

NOISE ATTENUATORS

Air-conditioning units produce noise at certain frequencies. Noise is transmitted both by air and solid structures. Attenuators can be employed in order to reduce the noise levels. Attenuators are generally made by a duct in galvanised sheet steel, which holds the splitters made of noise insulating material (generally mineral fiber panels). These splitters are placed in sheet elements and separated from the air by a particular fabric (fiber glass). Attenuators cause a certain pressure drop, which must be taken into account when setting the operating conditions of the air-conditioning system.

3.3.3 Cold storages

General Description

In cooling systems using chilled water as the cooling medium, a cold storage component can be integrated. This allows the chiller to be operated at times other than when cooling in the rooms is needed. In conventional cooling systems, storage has become one of the primary solutions to overcome the electric power imbalance between daytime demand and nocturnal abundance. The storage unit uses off-peak power to provide cooling capacity by extracting heat from a storage medium. Typically, a storage system uses refrigeration equipment at night to operate more efficiently than during the day, reducing the power consumption, and to create a reservoir of cold material. During the day, the reservoir is tapped to provide cooling power.

In a solar-assisted air-conditioning system, the reason to install a storage unit is somewhat different. The main purpose is to increase the use of solar energy in order to overcome periods of low radiation, in which the solar heat is not sufficient to cover the cooling load. This is achieved by generating more cooling power than needed during periods of high solar gains. Different time scales are of importance:

- Shortages of the solar radiation on time scales of seconds to minutes are due to clouds; if heat cannot be stored, the cooling process has to follow the available solar heat and in phases when the load exceeds the generated cooling power, a reservoir to allow further cooling is needed. In such cases, the cold reservoir has to be designed to cover the load only for a short time.
- Mismatches between solar gains and loads on a diurnal level can occur due to typical load patterns with a peak of the required cooling in the afternoon or evening, whereas the solar gains have their peak at noon. In such cases, the storage system has to be designed to cover the load on the order of hours.

Three Major Types

Three main types of storage systems are usually used in cooling systems:

- Eutectic salt
- Ice
- Chilled water

Eutectic Salt Storage

An eutectic salt/water solution is the storage medium using a phase change for storage purpose. Eutectic salt systems use a combination of inorganic salts, water and other elements to create a mixture that freezes at a desired temperature (typically near 8°C). The material is encapsulated in plastic containers that are stacked in a storage tank, through which water is circulated.

An advantage of these components is that smaller tanks are required than for chilled water, and by freezing at 8°C, standard chillers producing 5°C chilled water in commercial facilities can be used.

A drawback is that the tank cools the water for the distribution system to only 8 - 10°C, which achieves less dehumidification of the building and requires more pumping energy. That in turn limits the operating strategies that may be applied. For example, eutectic salts may only be used in operation relying entirely on storage if dehumidification requirements are low.

Ice Storage

Ice-based thermal storage systems use the latent heat of fusion of water - 335 kJ/kg - to store cooling capacity. To store energy at the temperature of ice requires refrigeration equipment that provides charging fluids at temperatures below the normal operation range of conventional air-conditioning equipment. They differ in how the cold from the ice is distributed to the building: conventionally, as cold air or rooftop.

One of the advantages of this type of storage is the compact size which amounts to 10 to 20 % of a comparable chilled water tank, and to 30 to 50 % of a comparable tank using an eutectic salt/water as storage medium. With cold air production, the distribution system (fans and ducts) can be down-sized. Other major benefits, when used with cold air or rooftop distribution systems, are the additional dehumidification effects and fan energy saving. An ice storage system can reduce the chilled water flow requirements by half.

A negative aspect is that most conventional chillers cannot be used. Then special ice-making equipment or standard chillers designed for low-temperature service are used. Electrical reciprocating, screw scroll or multi-stage centrifugal chillers and gas reciprocating chillers are normally used to produce ice.

CONSTRUCTION TYPES

Several technologies are available for charging and discharging ice storage units:

- Ice-harvesting systems feature an evaporator surface on which ice is formed and periodically released into a storage tank that is partially filled with water.
- External melt ice-on-coil systems use submerged pipes through which a refrigerant or secondary coolant is circulated. Ice accumulates on the outside of the pipes. The storage unit is discharged by circulating the warm return water over the pipes, melting the ice from the outside.
- Internal melt ice-on-coil systems also feature submerged pipes on which ice is formed. Storage is discharged by circulating warm coolant through the pipes, melting the ice from the inside. The cold coolant is then pumped through the building cooling system or used to cool a secondary coolant that flows through the building's cooling system.
- Encapsulated ice systems use water inside submerged plastic containers that freezes and melts as cold or warm coolant is circulated through the storage tank holding the containers.
- Ice slurry systems store water or water/glycol solutions in a slurry state, i.e., a partially frozen mixture of liquid and ice crystals. To meet a cooling demand, the slurry may be pumped directly to the load or to a heat exchanger cooling a secondary fluid that circulates through the building's chilled water system.

Internal melt ice-on-coil systems are the most commonly used type of ice storage technology in commercial applications. External melt and ice-harvesting systems are more common in industrial applications, although they can also be applied in commercial buildings and district cooling systems. Encapsulated ice systems are also suitable for many commercial applications. Ice slurry systems have not been widely used in air-conditioning applications.

CHILLER WATER STORAGE

Chilled water storage systems use the sensible heat capacity of water to store cooling energy. They operate at temperature ranges compatible with standard chiller systems and are most economical for systems greater than 7000 kWh in capacity. The storage energy density of a chilled water storage unit is given by the heat capacity of water:

$$q_{chilled,\, water} = \rho_{water}\, c_{water}\, \Delta T \approx 1.16 \left[\frac{kWh}{m^3 K} \right] \Delta T \qquad (3.12)$$

Figure 3.25
Examples of chilled water tanks made of steel (top), concrete (center) and plastic material (bottom).

For a typical chilled-water distribution system with a supply temperature of 6°C and a return temperature of 12°C, the useful temperature difference is 6°C. In such systems the storage density is about 7 kWh/m^3. However, in some systems much higher temperatures of the chilled water are sufficient; in general this holds if no air dehumidification has to be realised using the chilled water (e.g., chilled ceilings, hybrid desiccant cooling unit with cooler supplied with chilled water). Then the useful temperature difference may increase to a range of 12°C and the storage density increases to about 14 kWh/m^3. The dependence of the storage density related to the pure storage material, i.e., without insulation, on the usable temperature difference is shown in Figure 4.17 (see Section 4.2.1).

Advantages and Disadvantages

The advantages of storing cold in comparison to the storage of heat in a solar-assisted air-conditioning system are:

- The amount of energy to be stored to cover a certain load is lower for a cold medium than for heat if the COP is less than 1.0 as it is the case for most thermally driven chillers. This fact is also demonstrated in Figure 4.17 (see Section 4.2.1).

- The temperature of chilled water is generally closer to the temperature of the air surrounding the storage unit. This causes lower heat losses for the same standard of insulation.

A major disadvantage is that the storage density is lower because of the lower useful temperature difference. Chilled water storages require much larger storage tanks than the other storage media. Also the insulation quality has to be higher in comparison to hot water storage since condensation on the tank walls and particularly on the pipe connections has to be prevented.

Storage tanks must have the strength to withstand the pressure of the storage medium, and must be watertight and resistant to corrosion. Outdoor tanks must be weather-resistant. Examples for the different types of chilled water tanks are shown in Figure 3.25. Options for tank materials are:

Construction Types

- *Steel*: large steel tanks, holding up to several thousand cubic metres, are typically cylindrical in shape and constructed of welded plate steel on site. Some kind of corrosion protection such as an epoxy coating is usually required to protect the tank interior. Small tanks, with capacities of less than 100 m^3, are usually rectangular in shape and typically made of galvanised sheet steel. Cylindrical pressurised tanks are generally used to hold between 10 m^3 and 250 m^3.

- *Concrete*: concrete tanks may be pre-cast or cast-in-place. Pre-cast tanks are most economical in sizes of 1000 m^3 or more. Cast-in-place tanks can often be integrated into building foundations to reduce costs. However, cast-in-place tanks are more sensitive to thermal shock. Large tanks are usually cylindrical in shape, while smaller tanks may be rectangular or cylindrical.

- *Plastic*: plastic tanks are typically delivered as prefabricated modular units. UV stabilisers or an opaque covering are required for plastic tanks used outdoors to protect them against the ultraviolet radiation in sunlight.

4 THE HEAT PRODUCTION SUB-SYSTEM

The heat production sub-system is the part of the system which provides heat to a thermally driven air-conditioning system. With regard to the solar-driven equipment the solar collector is the main component of the sub-system. Besides the solar collector field other key components such as the thermal storage unit are considered. In addition, a back-up heat source can be present in solar-assisted systems, depending on system needs. The key components of the heat production sub-system are described in the following sections.

4.1 Solar collectors and back-up heat source

The solar collector is the main component to convert solar energy to the thermal energy that drives a solar-assisted air-conditioning system. In this section, the main solar collector types and their specifications are presented and discussed. Several terms related to solar geometry and basic concepts referring to solar radiation are used, but not explained. To deepen knowledge on these matters, the reader is referred to specific textbooks (for example /4.1/).

GENERAL CONSTRUCTION

The central component in each solar collector is the absorber. Here, the absorbed solar radiation is transformed in heat; part of this heat is transferred to the heat transfer fluid and another part is lost to the environment. Additionally, every collector - except uncovered collectors, which are used e.g., for swimming pool heating - contains a transparent cover, which separates the absorber from the environment and simultaneously transmits as much incident solar radiation as possible. For some collectors, opaque insulation is another important component.

COLLECTOR ENERGY BALANCE

In a steady state, the incident radiation on the collector surface is equal to the sum of useful heat and several different loss terms, as is evident from the energy balance of the absorber of a solar thermal collector which is given by the following expression:

$$A \cdot G_{\perp} = \dot{Q}_{use} + \dot{Q}_{loss,\ opt} + \dot{Q}_{loss,\ convective} + \dot{Q}_{loss,\ conductive} + \dot{Q}_{loss,\ radiative} \qquad (4.1)$$

HEAT LOSS MECHANISMS

where A is the absorber area in m² and $G_{\perp}$ is the incident global (total) solar radiation on the collector surface in W/m². $\dot{Q}_{use}$ is the useful heating power of the collector in W, and $\dot{Q}_{loss,\ opt}$ denotes the optical collector losses in W; this term includes all losses that are due to reflection and absorption in the transparent cover, i.e., the part of the incident radiation that does not reach the absorber. $\dot{Q}_{loss,\ convective}$ expresses convective losses in W; natural convection occurs in the gap between the absorber and the transparent cover and creates a heat flux from the absorber to the cover. $\dot{Q}_{loss,\ conductive}$ expresses heat conduction losses in W; they occur in the gap between the absorber and the cover as well as through the back insulation (depending on the construction of the collector) and through the frame edges. $\dot{Q}_{loss,\ radiative}$ denotes radiative losses in W; because of the absorber temperature these losses occur mainly in the infrared region of the spectral range.

REDUCTION OF HEAT LOSSES

The diverse losses contribute differently to the steady state energy balance, depending on the operating temperature of the collector and thus the temperature of the absorber. Different measures to reduce these losses can be taken in order to maximise the useful heat output from the solar collector. Selection of the most effective measures depends on the desired operating temperature and the construction of the collector. The measures that can be implemented alone, or in combination, include:

- Minimisation of transmission losses by applying an anti-reflective coating to the transparent cover. Using this kind of cover glass, the solar transmittance can exceed up to 96 %; an increase of the annual collector output in the range of 5-8 % can be achieved /4.2/.
- Minimisation of convective losses by evacuating the gap space between the absorber and the transparent cover. Depending on the geometry of the collector, a reduction of the pressure to some 100 Pa is sufficient to completely suppress natural convection. Other solutions are the use of a double cover (two glass panes or one glass pane and a polycarbonate film) or the use of a transparent insulation material (TIM) in the form of a honeycomb, which prevents the formation of large convection cells and, in this way, reduces convective losses.
- Minimisation of conductive losses. Depending on the type of collector, two main measures are applicable: for collectors with opaque insulation on the back, increasing the insulation thickness as well as use of a material with lower conductivity results in lower conductive losses. But also the gas in the gap between the absorber and transparent cover contributes to heat conduction. Here, either filling with an inert gas or again evacuation leads to an improvement. However, for a notable reduction of heat conduction in the gap, evacuation to a final pressure of a few Pa is necessary. This can only be achieved using very gas tight constructions and compounds able to absorb gas molecules.
- Minimisation of radiative losses can be achieved by use of selective coatings on the absorber. These coatings have a high absorptance in the visible range of the solar spectrum but low emittance in the infrared range. Until a few years ago, only coatings produced on the basis of electro-chemical processes were available. Today most coatings are produced in more environmentally friendly ways such as sputtering or physical vapour deposition (PVD).
- Finally reduction of all thermal losses can be achieved by optical concentration. The useful radiation on the collector is proportional to the aperture area of the collector, while all thermal losses are proportional to the absorber area (for areas definition see below). Therefore a construction in which the aperture area is larger than the absorber area leads to a reduction of thermal losses. One approach is to use mirrors, which concentrate the radiation on the absorber. However, geometrical concentration is only possible for direct radiation. Therefore systems with high radiation concentration factors have to track the sun, whereby part of the diffuse radiation is lost. The concentration factor is given by the ratio of the aperture area to the absorber area (see Section 4.1.3).

Collector Efficiency

A measure of collector performance is the collector efficiency. The solar collector efficiency is defined as the ratio of the useful heat output to the total global radiation incident on the collector surface:

$$\eta = \frac{\dot{Q}_{use}}{A \cdot G_{\perp}} \qquad (4.2)$$

The efficiency of a solar collector can be written as follows:

$$\eta = k(\Theta) \cdot c_0 - c_1 \cdot \frac{(T_{av} - T_{amb})}{G_{\perp}} - c_2 \cdot \frac{(T_{av} - T_{amb})^2}{G_{\perp}} \qquad (4.3)$$

This equation follows directly from the energy balance in Equation 4.1 and the definition of the efficiency factor given in Equation 4.2, if all non-linear losses are approximated by a quadratic expression. The symbols used in Equation 4.3 have the following meaning: $k(\Theta)$ is the incident angle modifier, which accounts for the influence of non-perpendicular incident radiation at incidence angle, Θ, in relation to normal incidence radiation Θ = 0. The expression above represents only an approximation for the influence of the incident angle on the optical performance of the solar collector. In a detailed physical description the impact of the incident angle has to be considered separately for direct and diffuse radiation. Examples of the incident angle modifier for different collector types are given in the sections below. T_{av} is the average fluid temperature in the collector; for typical flowrates, the average fluid temperature can be expressed by the arithmetic average between the inlet and the outlet temperature. T_{amb} is the ambient air temperature. c_0, c_1 and c_2 are the collector efficiency values, whereby c_0 denotes the optical efficiency value and c_1 the linear and c_2 the quadratic loss coefficients, respectively.

Collector Area Definition

Different definitions of the solar collector area are possible; they are shown using the example of a flat-plate collector in Figure 4.1:

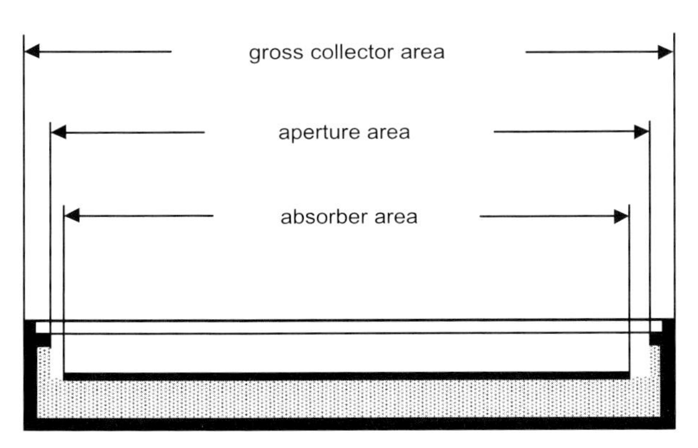

Figure 4.1
Definition of the different collector areas for a flat-plate collector.

- the gross collector area is defined by the external dimensions of the collector,
- the aperture area is defined by the geometry of the transparent cover or by the projection area of the reflector in the case of concentrating collectors, and
- the absorber area is the effective size of the radiation-absorbing component.

The collector efficiency values can vary considerably depending on which of these areas is used as a reference (see Table 4.1 in Section 4.1.5). Therefore, it is very important to indicate precisely which reference area is used in the calculations.

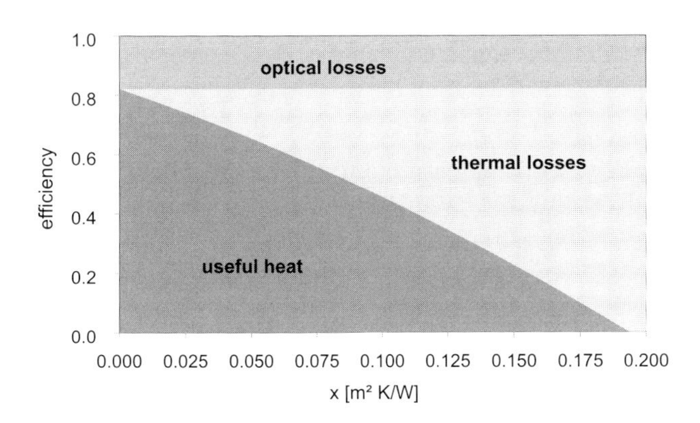

Figure 4.2
Typical solar collector efficiency curve. The efficiency is expressed as a function of parameter x, defined in Equation 4.4; the major losses and useful gains are marked.

Collector Efficiency

In this book, the absorber area is implied always if not indicated otherwise. The collector efficiency curves are usually expressed as a function of the ratio of the difference between the average fluid temperature, T_{av}, and ambient air temperature, T_{amb}, to the incident global solar radiation on the collector, $G_\perp$, as follows:

$$\eta = k(\Theta) \cdot c_0 - c_1 \cdot x - c_2 \cdot x^2 \cdot G_\perp \quad \text{where:} \quad x = \frac{(T_{av} - T_{amb})}{G_\perp} \qquad (4.4)$$

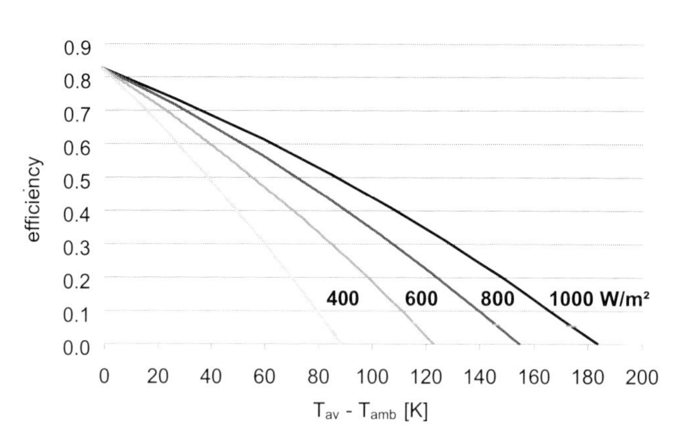

A general efficiency curve, identifying different collector losses, is shown in Figure 4.2. Another common presentation of the collector efficiency is illustrated in Figure 4.3. In this case, the solar collector efficiency curves are plotted for different values of the incident solar radiation on the collector as a function of the temperature difference between fluid and ambient air, T_{av}-T_{amb}. Figure 4.3 demonstrates that for a certain temperature difference, the efficiency decreases with decreasing solar radiation; thus the energy gains of a solar thermal collector will be reduced more than proportionally to the solar radiation.

4.1.1 Flat-plate collectors [1)]

MARKET

Flat-plate collectors are the most common solar collector type and they dominate the market. In all countries, except China, they represent about 90 % of the market of covered solar collectors. The most popular application is for domestic hot water production. During the year 2000 alone, the total newly installed area of flat-plate solar collectors in the IEA member countries is estimated to be about 2.2 million square metres /4.3/.

COLLECTOR CONSTRUCTION

A flat-plate collector consists of a metallic absorber and a glass cover. The absorber consists either of several metal fins or a single absorber plate. One of the most critical issues that affects the performance of flat-plate collectors is the coupling of the tubes, where the heat transfer fluid circulates, to the absorber. Different types of manufacturing technology, such as laser or ultrasonic welding processes, are used for this purpose. The upper surface of the absorber dominates the thermal performance due to its optical characteristics.

SELECTIVE COATING

Most collectors used in temperate climates have a selective coating, which has a high absorptance in the visible range of the solar spectrum, α_{vis}, but a low emissivity, ε_{IR}, in the infrared range. Selective coatings are characterised by an emissivity ε_{IR} below 0.2 while non-selective surfaces have an emissivity ε_{IR} between 0.5 and 1. Both surfaces in general have an absorptance α_{vis} higher than 0.9. A schematic cross-section of a flat-plate solar collector and a photograph of a collector used in combination with a desiccant cooling system are shown in Figure 4.4.

[1)] *In general any collector using a flat geometry, i.e., a flat absorber and cover, is a flat-plate collector, independent of the heat fluid (liquid, air). However, the expression flat-plate collector usually refers to collectors with a liquid fluid. Therefore in this handbook, 'flat-plate collector' denotes a collector with a flat absorber cooled by a liquid heat transfer fluid. A collector which heats air is called a solar air collector.*

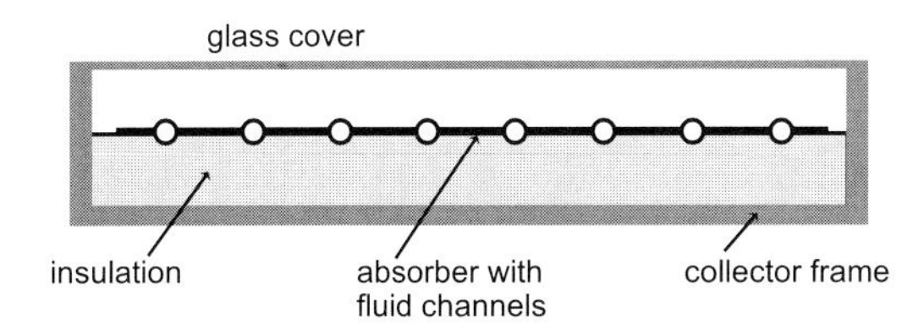

Figure 4.4
Schematic cross-section of a flat-plate collector (top) and photograph of a flat-plate collector system which provides heat to a desiccant cooling system in the Netherlands (bottom).

Flat-plate collectors are available in modules ranging from about 1.8 m^2 up to 10 m^2. The general trend is to manufacture larger modules, particularly for applications with large overall areas. Depending on the application, the solar collectors are installed on a simple supporting structure to provide optimum tilt and orientation, but they may also be integrated into a sloped roof, which is advantageous from an architectural point of view.

COLLECTOR EFFICIENCY

Plots of typical efficiency curves are shown in Figure 4.5. Typical examples of characteristic values for flat-plate collectors are given in Table 4.1 (in Section 4.1.5). Beside the efficiency values c_0, c_1 and c_2 also values of K_{50} are shown in Table 4.1. The latter are incident angle modifier (IAM) values and they account for the reduction of the optical efficiency for an incidence angle $\Theta = 50°$ in comparison to the solar radiation normally incident on the collector, $\Theta = 0°$. A typical pattern of the IAM value´s dependence on the incidence angle is described in Figure 4.6. It shows that the IAM remains close to 1 for incidence angles smaller than 60°.

Table 4.1 also provides exemplary cost values for flat-plate solar collectors. These values are for the collector only and do not include the supporting structure or the hydraulic system. In addition no value added tax (VAT) is included either. Due to their performance characteristics, selective flat-plate collectors can be used in combination with different cooling techniques for solar-assisted air-conditioning (see Figure 4.13). They are well suited for use in combination with desiccant cooling systems and for chilled water production using adsorption chillers and eventually single-effect absorption chillers. However, only high-quality collectors with a selective absorber coating are suitable because of the typical driving temperatures for this kinds of cooling technique.

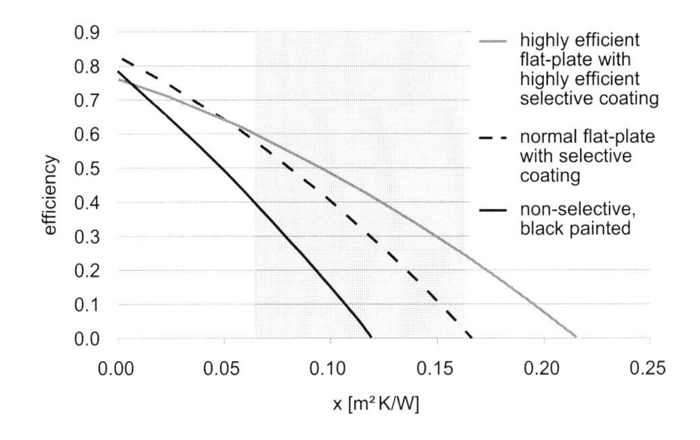

Figure 4.5
Typical efficiency curves for different types of flat-plate collectors; the grey shaded area represents the typical application range for an absorption chiller (ambient air temperature 30°C, collector average temperature 95°C, radiation on collector from 400 W/m², which corresponds to the right end of the shaded area, x=0.1625, to 1000 W/m², which corresponds to the left end of the marked area, x=0.065).

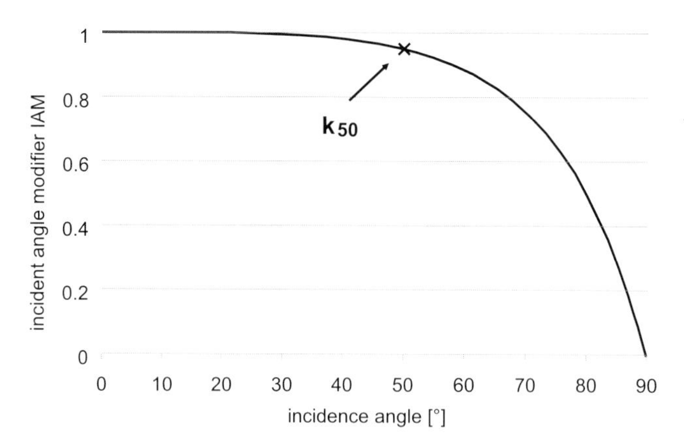

Figure 4.6
Incident angle modifier of a typical flat-plate collector; K_{50} indicates the reduction of optical efficiency at an incidence angle of 50°. The IAM at normal incidence radiation is 1.

4.1.2 Evacuated tube collectors

MARKET

Evacuated tube collectors represent about 10 % of the market of covered solar collectors in IEA member countries /4.3/. In China, this collector technology is the dominating one and the sales volume in 2001 was several million square metres. An evacuated tube collector always consists of single tubes which are connected to a header pipe. Each single tube is evacuated in order to reduce heat losses. The tubular geometry is necessary in order to withstand the pressure difference between the atmospheric pressure and the internal vacuum (typically a few Pa). Evacuated tube collectors can be grouped into two main categories:

COLLECTOR CONSTRUCTION

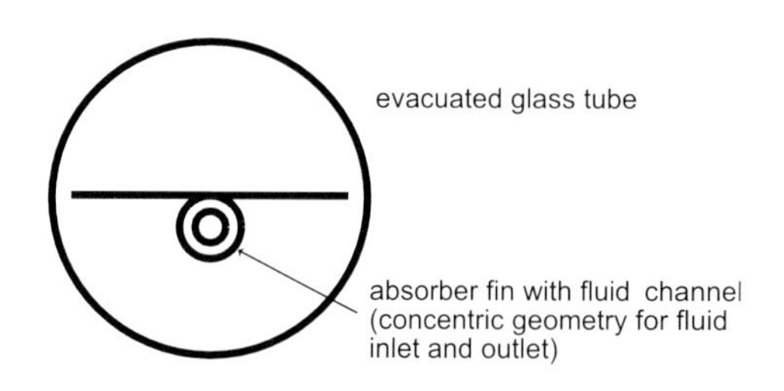

Figure 4.7
Schematic cross-section of an evacuated tube collector (top) and photograph of a collector array, which supplies heat to a solar-assisted adsorption chiller in a laboratory in Germany.

- Tubes with flow of the heat transfer fluid through the absorber, and
- tubes with heat transfer between the absorber and heat transfer fluid of the collector cycle using the heat-pipe principle.

There are three main geometrical configurations for the types with direct contact:

- Systems with concentric fluid inlet and outlet: the advantage of this construction is that the connection to the header pipe is symmetric to rotation; therefore, each single pipe can be easily rotated, allowing the absorber fin to have the needed tilt angle even if the collector is mounted horizontally.
- Systems with two separate pipes for inlet and outlet: this is the traditional evacuated tube collector construction.
- 'Sydney' type collector: this collector consists of a Dewar-like double glass tube with an integrated cylindrical metal absorber. The advantage is that the boundary between the vacuum and the atmosphere is all-glass, which reduces sealing problems or outgassing of materials in the vacuum.

In an evacuated tube collector of the heat-pipe type a 'dry' connection is made between the absorber and the header; this makes the installation process easier. However, the collector must always be mounted with a certain tilt angle in order to allow the condensed internal fluid of the heat pipe to return to the hot absorber.

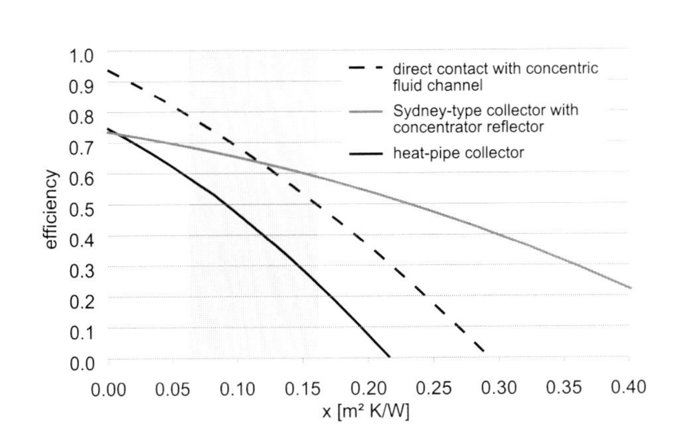

Figure 4.8
Efficiency curves for typical evacuated tube collectors; the grey shaded area indicates a typical application range for an absorption chiller (ambient air temperature 30°C, collector average temperature 95°C, radiation on collector from 400 W/m² which corresponds to the right end of the marked area, x=0.1625, to 1000 W/m² which corresponds to the left end of the marked area, x=0.065).

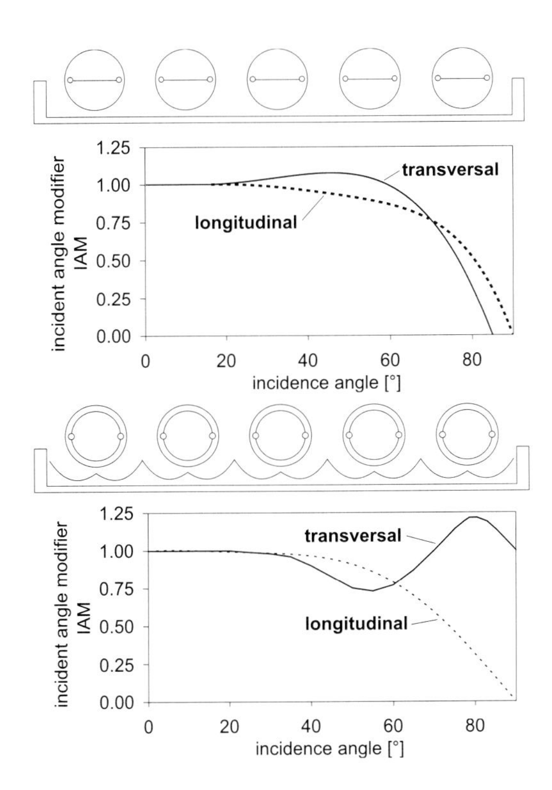

Figure 4.9
Incident angle modifiers of two different evacuated tube collectors. Evacuated tube collector with flat absorber and without reflector (top), evacuated tube collector with cylindrical absorber and CPC reflector (bottom). The longitudinal IAM applies for beams parallel to the tube axis and the transversal IAM for beams perpendicular to the tube axis.

Sometimes reflectors are mounted behind evacuated tube collectors. This leads to increased use of the radiation on the collector since radiation is absorbed which otherwise would pass through the gap between two neighbouring tubes, and would not reach the absorber. Also systems with reflectors which are constructed for concentration of the incoming direct radiation are available.

COLLECTOR EFFICIENCY

A schematic drawing of an evacuated tube collector with direct contact between the absorber and the heat transfer fluid is shown in Figure 4.7. The figure also includes a photo of an evacuated tube collector array, which supplies heat to an adsorption chiller for air-conditioning of a laboratory in Germany. Examples of the efficiency values of some different types of evacuated tube collectors are given in Table 4.1 (Section 4.1.5.). Typical efficiency curves of evacuated tube collectors are shown in Figure 4.8.

USE FOR SOLAR-ASSISTED AIR-CONDITIONING

For evacuated tube collectors, it is necessary to distinguish the influence of the incidence angle of the incoming solar radiation in different directions. Therefore two values, K_{long} and K_{trans}, are given in Table 4.1, which correspond to planes of incidence parallel and perpendicular to the tube axes, respectively. Typical IAM curves are shown in Figure 4.9. Exemplary cost data for different examples are given in Table 4.1 in Section 4.1.5.

4.1.3 Stationary CPC collectors

Concentrating solar radiation on a relatively small collection area can improve the collector performance. Collectors using a stationary compound parabolic concentrator (CPC) can achieve a small solar radiation concentration reaching a higher efficiency without significantly increasing the production and operation costs compared to flat-plate collectors.

OPTICAL CONCENTRATION

This type of collector is characterised by the concentration ratio, C, which is the ratio of the collector aperture area, $A_{aperture}$, to the collector absorber area, $A_{absorber}$:

$$C = \frac{A_{aperture}}{A_{absorber}} \tag{4.5}$$

The acceptance angle, Θ_a, also characterises a concentrating collector and is defined as the maximum incidence angle of the beam (direct) solar radiation on the collector aperture that will reach the absorber without tracking. Considering the above definitions of concentration ratio and acceptance angle, it is clear that small acceptance angles are related to high concentration factors and vice versa. For a certain acceptance angle, there is a maximum concentration factor that a collector can achieve. This maximum concentration is imposed by fundamental physical principles. For two-dimensional geometry, this maximum concentration, C_{max} is given by:

$$C_{max} = \frac{1}{\sin \Theta_a} \tag{4.6}$$

Collectors that achieve this maximum concentration factor are considered 'ideal' concentrators. CPC collectors are 'ideal' concentrators. They are versatile because they can be designed to accommodate different absorber shapes, for example, horizontal and vertical flat absorbers, tubular absorbers etc.. For each absorber shape the reflector shape is designed to ensure that all beam (direct) solar radiation reaching the collector aperture within the acceptance angle will reach the absorber, even if it is reflected more than once by the mirror (reflector).

Stationary CPC collectors used for heating of a liquid fluid at low (50-70ºC) and medium temperatures (80-110ºC), have two orthogonal axes symmetry and are designed with acceptance angles greater than 30º to avoid having to track the sun (moving the collector to follow the daily path of the sun). This means that they usually have concentration factors lower than 2.

Practical design limitations, such as the height of the collector box, can also impose the use of truncated CPC's. This means that the upper part of the mirror is cut and sunrays with incident angles higher than the acceptance angle can still impinge directly on the absorber, but when they strike the mirror, they are reflected out of the collector. A truncated CPC has a concentration factor which is determined by Equation 4.5 but is lower than the characteristic maximum concentration.

An important aspect of CPC's is that they collect beam, diffuse sky and ground-reflected solar radiation. This has to be taken into account when determining the collected solar energy for the different collector trough orientations, typically east-west for forced circulation systems or north-south for thermosiphon systems.

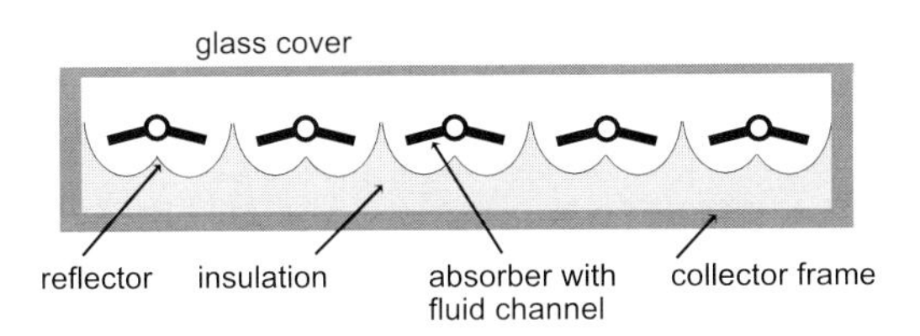

Figure 4.10
Schematic cross-section of a stationary CPC collector (top) and an example of a solar collector array installed on the south facade of the office building of a company in Portugal for driving a desiccant cooling system (bottom).

COLLECTOR EFFICIENCY

The efficiency of stationary CPC collectors is determined using the same standards as for flat-plate collectors and is characterised by the same Equation 4.4. The efficiency will depend on the collector's design and especially on the concentration ratio.

Manufacturers of stationary CPC collectors use selective surfaces for the absorbers, which means that these collectors can have lower heat losses than flat-plate selective collectors, depending on the concentration ratio they achieve.

A cross-section scheme of a stationary CPC collector is shown in Figure 4.10; the figure also shows an example of an installation where this type of collector is installed in combination with a desiccant cooling system. Such a collector is designed with concentration factors of about 1.12 for domestic hot water production and about 1.5 for production of heat in the range between 80°C and 110°C, that can be used by thermally driven chillers.

Exemplary values of the performance parameters and cost of a stationary CPC collector which is equipped with a teflon film between absorber and cover reducing convection losses are given in Table 4.1 in Section 4.1.5.

4.1.4 Solar air collectors

MARKET

In all applications where warm or hot air is needed the use of solar air collectors is a viable alternative. The main operating principles are similar as for solar collectors with liquid fluid but in this case an electric fan is used to circulate air through the collector. In Europe, the market for solar air collectors represents only 1 to 2 % of the solar liquid collector market. This is on the one hand due to the lack of knowledge among end users about the available technology and possible applications, and on the other hand the lack of experience with the design of solar air collector systems. Furthermore, solar air collectors cannot be used directly for domestic hot water production, which today dominates the market for solar collectors.

Typical applications for solar air collectors are heating of residential and non-residential buildings, and industrial processes where large flowrates of heated air are required. As far as solar-driven air-conditioning is concerned, they can be coupled with desiccant cooling technology.

ADVANTAGES AND DISADVANTES

The general advantages of solar air collectors are:

- No freezing problems during winter,
- No overheating problem in summer,
- Simple system components,
- No water leakage.

The general disadvantages of solar air collectors are:

- No standard heat storage units available on the market.
- Commonly the fan electricity consumption due to pressure drops is higher than for pumps in a solar system employing liquid-based collectors and having the same collector array dimensions.
- Lower collector efficiency than liquid-based solar collectors due to lower heat transfer rate from the absorber to the air-flow.

Collector Construction

The general construction of solar air collectors is similar to that of liquid-based solar collectors. The commercially available solar air collectors include: covered collectors and unglazed collectors.

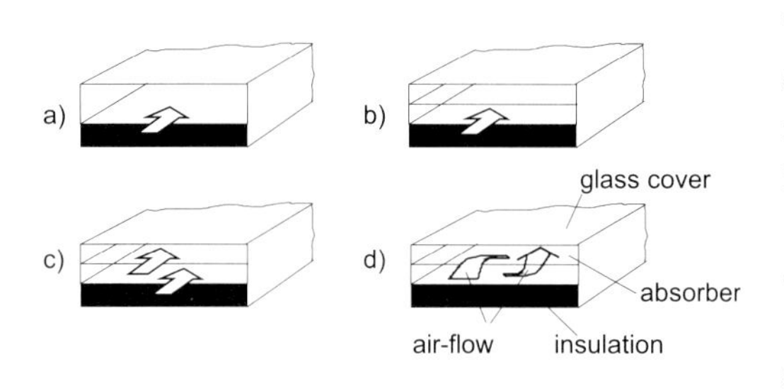

Figure 4.11
Schematic drawing of different types of flat-plate solar air collectors.

The covered collectors may be divided into four categories, which differ in the way that the air-flows relative to the absorber, (Figure 4.11) as follows:

a) air-flow above the absorber

b) air-flow below the absorber

c) air-flow around the absorber

d) air-flow through the perforated absorber.

Figure 4.12
Schematic cross-section of a solar air collector with airflow below the absorber (top); an example of a solar air collector array which is used for regeneration of the desiccant rotor in a desiccant system for air-conditioning of a seminar room in Germany (bottom).

The type (a) collector has the simplest construction, since the absorber is positioned on top of the insulation. The main disadvantage is the lower efficiency of the collector, since the heated air-flow is in direct thermal contact with the cold glass cover, resulting in high thermal losses. For type (b) these heat losses are significantly reduced as the air-flow is below the absorber. To improve the heat transfer from the absorber to the air-flow, the absorber is usually produced with fins. For type (c), the heat transfer to the air-flow is even better but again there is an increase of thermal heat losses towards the cold glass cover. Type (d) is the one with the highest heat transfer rate, but on the other hand, the pressure drop in the collector is higher, which results in higher electricity consumption by the fan to force the air-flow through the collector.

A schematic cross-section of a solar air collector is shown in Figure 4.12. The figure also shows an example of a solar air collector, which is used for supplying hot air to regenerate a desiccant rotor in a desiccant cooling system for air-conditioning of a seminar room in Germany.

In selecting the best solar air collector construction, not only the thermal efficiency but also the electricity consumption for moving the air through the collector must be taken into account. Type (b) is the one with the best overall performance and as a result, it has the highest market share.

EFFICIENCY

The efficiency curve of a solar air collector is usually given in the same form as for solar liquid collectors. The main difference is that the temperature difference that appears in Equation 4.3 or 4.4 is not (T_{av} - T_{amb}) but is defined as the difference between the outlet air temperature, T_{out}, and the ambient air temperature, T_{amb}, i.e., (T_{out} - T_{amb}). The reason for using the collector outlet air temperature instead of the average collector temperature is the higher temperature increase in a solar air collector compared to a solar liquid collector. This causes a non-linear temperature increase in the collector, so that the outlet temperature is more representative for the collector temperature than the average of the inlet and outlet temperatures.

The costs of solar air collectors lie in the range of about 200 - 430 €/m², depending on the collector array dimensions. These cost figures include the supporting construction.

4.1.5 Summary - solar collectors

WORKING RANGES

Several different types of solar collector can be used in air-conditioning systems, depending on the application (see Table 4.1 and Figure 4.13). In Table 4.1 examples of the most important types of non-tracking solar collectors using a liquid fluid are given. Efficiency and cost parameters are shown with reference to different collector areas (absorber, aperture, gross area). The main parameters that have to be specified are the temperature ranges for the different types of thermally driven cooling technology and the performance of the various solar collector types (Figure 4.13). However, it should be noted that the efficiency curve cannot provide sufficient information for the selection of a certain collector for a certain application; the operation ranges shown in Figure 4.13 have only to be considered as a rough indication.

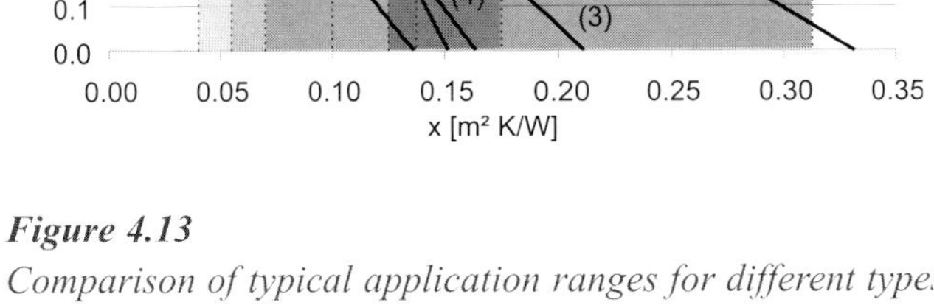

Figure 4.13
Comparison of typical application ranges for different types of solar collectors. The labels on the efficiency curves refer to the following collector types:

(1) highly efficient evacuated tube collector (SYDNEY type), (2) direct-contact evacuated tube collector, (3) heat-pipe evacuated tube collector, (4) typical flat-plate collector with selective coating, (5) solar air collector with selective coating, (6) stationary CPC collector with Teflon film for convection reduction. For each of the indicated application ranges, the left side of the coloured area represents a radiation on the collector of 1000 W/m², while the right side represents a radiation of 400 W/m². [The ambient air temperature was assumed to be 25°C.] The typical collector operation temperature (average fluid temperature) for the different cooling techniques is indicated in brackets: DEC desiccant cooling technique (operation, temperature 65°C), ADS adsorption chiller (80°C), 1-ABS single-effect absorption chiller (95°C), 2-ABS double-effect absorption chiller (150°C).

		Examples of evacuated tube collectors					
		parameters referring to			parameters referring to		
	Unit	absorber area	aperture area	gross area	absorber area	aperture area	gross area
Manufacturer	-	microterm Energietechnik GmbH			Thermomax Ltd.		
Name	-	Sydney SK-6			Memotron TMO 600		
Type	-	Evacuated tube collector, cylindrical absorber, directly cooled, CPC concentrator			Evacuated tube collector, flat absorber, heat-pipe		
Area of single module	m^2	0.984	1.088	1.181	1.975	2.154	2.762
c_0	-	0.735	0.665	0.612	0.84	0.770	0.601
c_1	W/m^2K	0.65	0.59	0.54	2.02	1.85	1.44
c_2	W/m^2K^2	0.0021	0.0019	0.0017	0.0046	0.0042	0.0033
$k_{1,long}$ (50°)	-		0.93			0.91	
$k_{2,trans}$ (50°)	-		0.79			0.96	
η (ΔT=50 K, 800 W/m^2)	-	68.8%	62.2%	57.3%	69.9%	64.1%	50.0%
η (ΔT=70 K, 800 W/m^2)	-	66.5%	60.2%	55.4%	63.5%	58.2%	45.4%
Specific cost	€/m^2	771	698	643	777	712	555
Reference	-		/4.4/			/4.4/	
		Examples of flat-plate collectors					
		parameters referring to			parameters referring to		
	Unit	absorber area	aperture area	gross area	absorber area	aperture area	gross area
Manufacturer	-	Sonnenkraft Vertriebs GmbH			Wagner & Co Solartechnik GmbH		
Name	-	SK 500			Euro C18		
Type	-	Flat-plate collector, selective coating			Flat-plate collector, selective coating		
Area of single module	m^2	2.215	2.307	2.567	2.305	2.34	2.612
c_0	-	0.8	0.768	0.690	0.789	0.777	0.696
c_1	W/m^2K	3.02	2.90	2.61	3.69	3.63	3.26
c_2	W/m^2K^2	0.0113	0.0108	0.0098	0.007	0.0069	0.0062
$k_{1,2}$ (50°)	-		0.94			0.91	
η (ΔT=50 K, 800 W/m^2)	-	57.6%	55.3%	49.7%	53.7%	52.8%	47.3%
η (ΔT=70 K, 800 W/m^2)	-	46.7%	44.8%	40.3%	42.3%	41.7%	37.4%
Specific cost	€/m^2	271	260	234	265	261	234
Reference	-		/4.4/			/4.4/	
		Example of a roof-integrated collector			Example of a stationary CPC collector		
		parameters referring to			parameters referring to		
	Unit	absorber area	aperture area	gross area	absorber area	aperture area	gross area
Manufacturer	-	Sun - Pro Gmbh			AO SOL, Lda		
Name	-	Sunbox HFK-S			CPC AO SOL 1.5		
Type	-	Flat-plate collector, selective coating, roof-integrated			Stationary CPC collector (concentration 1.5), selective coating, teflon film		
Area of single module	m^2	5.444	5.478	6.082	1.59	2.38	2.69
c_0	-	0.786	0.781	0.704	0.94	0.628	0.556
c_1	W/m^2K	3.38	3.36	3.03	2.2	1.47	1.30
c_2	W/m^2K^2	0.0107	0.0106	0.0096	0.033	0.0220	0.0195
$k_{1,2}$ (50°)	-		0.92			-	
η (ΔT=50 K, 800 W/m^2)	-	54.1%	53.8%	48.5%	69.9%	46.7%	41.3%
η (ΔT=70 K, 800 W/m^2)	-	42.5%	42.2%	38.0%	54.5%	36.4%	32.2%
Specific cost	€/m^2	196	195	175	377	252	223
Reference			/4.4/			/4.5/4.6/	

Table 4.1

Typical examples of characteristic values and cost for different solar collector typologies. Note: $k_{1,long}(50°)$ and $k_{2,trans}(50°)$ denote longitudinal and transversal incident angle modifiers of evacuated tube collectors at an incidence angle of 50°. $k_{1,2}(50°)$ denotes the incident angle modifier of the other collectors, where the orientation of the incidence angle has not to be taken into consideration. Efficiency values, η, are given for two different operation conditions. ΔT denotes the difference between the average fluid temperature in the collector and the ambient air temperature.

SOLAR COLLECTORS AND THERMALLY DRIVEN COOLING

More detailed methods for collector selection are presented in Chapter 6. The following guidelines give a very general indication:

- For double-effect absorption chillers, highly efficient evacuated tube collectors are the only suitable type among the ones presented here. Collectors with single-axis or dual-axis tracking could also be used, but they are not treated in this handbook.
- For single-effect absorption chillers not only all types of evacuated tube collectors can be considered but also stationary CPC collectors and possibly highly efficient flat-plate collectors.
- For adsorption chillers either flat-plate collectors, stationary CPC collectors or evacuated tube collectors can be used.
- For desiccant cooling, stationary CPC collectors, flat-plate collectors or solar solar air collectors are a good choice.

In all cases, a collector with a selectively coated absorber should be chosen in order to achieve sufficiently good performance during operating periods with reduced solar radiation.

COSTS

Specific costs of solar collector systems decrease with the installed collector area. Typical cost curves, which are based on experience gained in several systems with large collector areas, are shown in Figure 4.14.

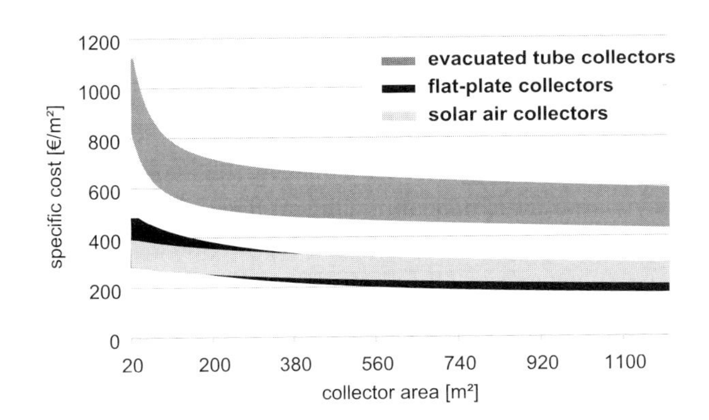

Figure 4.14
Cost ranges for different types of solar collectors: specific cost as a function of the collector area. The cost values include the supporting structure and piping of the collectors, but do not include the cost for heat storage unit and other system components (safety valves, expansion vessel, heat exchanger, heat transfer fluid, pumps etc.). Note that the costs referring to flat-plate collectors also apply to stationary CPC collectors.

4.1.6 Back-up heat source

There are various possible options to provide back-up in a solar-assisted air-conditioning application. Back-up can be employed on either the 'hot' or the 'cold' side. Back-up heat may be provided from different heat sources. In this handbook, it is generally assumed that gas heaters provide the thermal back-up.

CONSTRUCTION TYPES

Back-up heaters are needed to assure that solar-assisted air-conditioning systems always have enough heat available to meet the load. For the cooling process, they are needed especially during days with hot and humid weather and simultaneous clouds, i.e., on days with high cooling loads on which the solar system does not deliver enough heat at the required temperature. Additionally, they can be used for space heating during the heating season, when the heat output from the collectors is not sufficient. Gas heaters can be categorised according to different criteria, as follows:

- single-stage or capacity-controlled burner;
- atmospheric or ventilated burner;
- high-temperature, low-temperature and condensing burners;
- 'catalytic' burner;
- tank-integrated or separate boiler.

Figure 4.15 shows two examples of gas-fired water heaters. The heating capacity of gas-fired burners can have one or two fixed values (two-stage) or a range of values (typically from 30 % to 100 % nominal capacity). In the latter case, they are also referred to as modulating gas-fired burners. If the heating capacity is fixed, a water storage tank should also be coupled to the gas-fired burner in order to avoid cycling operation. Most of today's small and medium-sized gas-fired burners use atmospheric burners. This means that gas and air are mixed in a kind of venturi nozzle before they are ignited. The gas-air mixture burns upwards without a fan. The advantages are low noise emission and simple construction. Ventilated burners are more efficient due to the better mixing of air and gas, but are noisier and technically more complex.

ENERGY PERFORMANCE

The flue gas energy losses can be minimised by lowering the temperature of the flue gas of the burner using a heat sink. When the flue gas temperature drops below 60°C, the water content of the flue gas starts to condense. Below 30°C, most of the water has condensed. Since the condensing heat is also useful heat, the burner efficiency can increase up to about 110 %, based on the lower heating value of the fuel. Non-condensing burners reach efficiency values up to 95 %, based on the lower heating value of the fuel. In the event that condensation occurs, the burners must be protected against corrosion. For solar-assisted air-conditioning, the driving temperature is usually too high to achieve gas condensation.

When the combustion takes place around a special catalyst, the temperature can be lowered which results in lower NO_x emissions. Some manufacturers have included this type of burner in their gas-fired boilers.

Figure 4.15
Low temperature two-stage atmospheric burner (right) and modulating gas-condensing catalytic ventilated burner (left)(source: Viessmann).

Small gas-fired burners of up to 20 kW are sometimes directly mounted in the water storage tank. The boilers for burners with higher heating capacity are installed separately, because the heat transfer rate to the water of the tank is too low and the space available in the tank is too small.

In many systems the upper part of the water storage unit for the solar thermal collectors is also used as heat storage for the gas-fired boiler (see Figure 4.18 in Section 4.2.1 below). However, the gas-fired burner is assumed to be independent of the water storage tank.

4.2 Storage systems

COMMON APPLLICATION

The main purpose of storage in a solar-assisted air-conditioning system is to overcome mismatches between solar gains and cooling loads. The most common application is the integration of a hot water buffer tank in the heating cycle of the thermally driven cooling equipment. When a desiccant cooling system is used, this is the only place to integrate storage into the system.

However, in a solar-assisted air-conditioning system that uses absorption or adsorption chillers, there are two possible places for integrating thermal storage (Figure 4.16). One option is that the excess solar heat can be stored in the heat storage unit and is made available if the solar heat is not sufficient. The second option is that the excess cooling power of the solar system and the thermally driven chiller is stored in a cold storage unit and is made available if the cold production does not meet the cooling load (see Section 3.3.3).

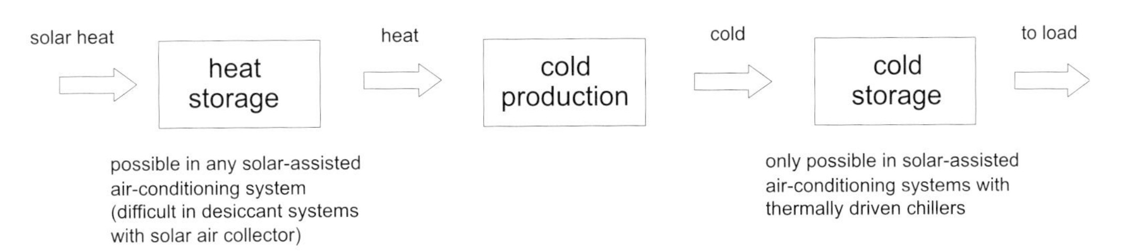

Figure 4.16
Energy storage typologies in a solar-assisted air-conditioning system.

The following sections describe the most common methods of heat storage in thermal energy systems and highlight the techniques that are most relevant in using solar thermal energy for air-conditioning.

4.2.1 Hot water storages

STORAGE FUNCTION

One of the key elements of a solar-assisted air-conditioning system is the hot water tank. The storage unit fulfils several tasks:

- It delivers sufficient energy to the heat sink (with appropriate mass flow and temperature).
- It decouples the mass flows between heat sources and heat sinks.
- It stores heat from fluctuating heat sources (i.e., solar) from times where excess heat is available for times where too little or no heat is available.
- It extends the operation times for auxiliary heating devices.
- It reduces the needed heating capacity of auxiliary heating devices.
- It stores the heat at the appropriate temperature levels avoiding mixing in order to reduce exergy losses (i.e., stratification).

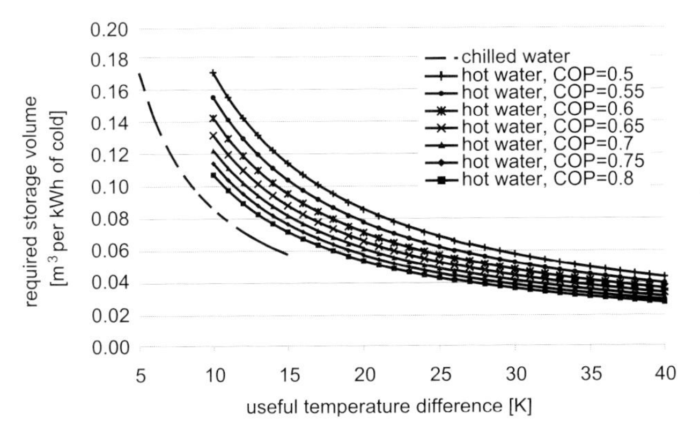

Figure 4.17
Required storage volume (without insulation) for chilled water storage and hot water storage for different values of the COP of a thermally driven chiller as a function of the useful temperature difference. The required storage volume is given in m^3 required to provide 1 kWh of cold.

Storage Volume

The minimum required hot water storage volume to provide cold to the load in a thermally driven cooling system is shown in Figure 4.17 in comparison to the respective figure for chilled-water storage.

Storage Construction

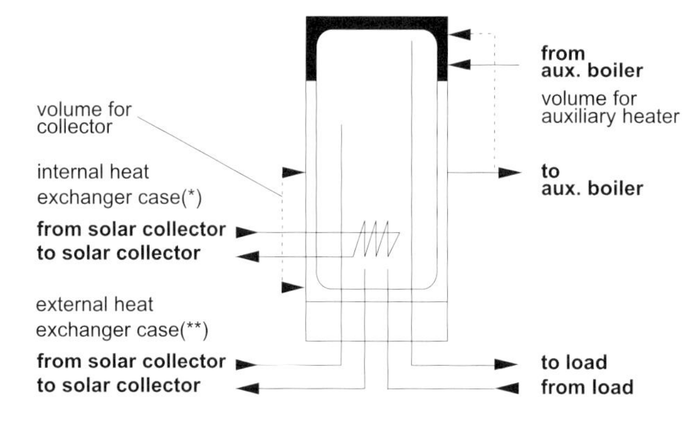

Figure 4.18

Schematic drawing of a hot water tank storage for solar-assisted air-conditioning system. Alternative constructions with either internal (*) or external heat exchanger (**) are shown.

Figure 4.18 illustrates the operating principles of a hot water storage tank with two heating inputs (solar and auxiliary); two alternative designs for coupling the solar collector system are shown, namely an internal heat exchanger and the connection to an external heat exchanger. The temperature of water flowing from the storage tank to the solar collectors should be as low as possible, in order to obtain the highest possible collector efficiency.

The configuration of the piping connections to the different heat sources and heat sinks are described next in order to demonstrate the complexity of such a system.

Internal Heat Exchanger

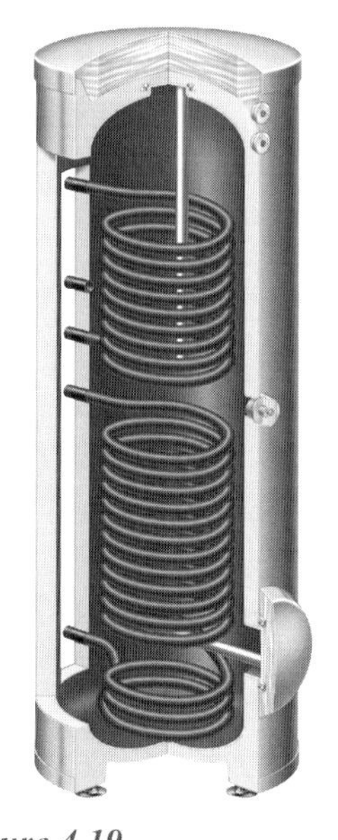

Figure 4.19

Hot water tank with internal heat exchangers for collector areas up to 15 m² (Source: Viessmann, Germany).

For small collector areas (smaller than 15 m²), an internal heat exchanger located at the bottom of the storage tank and a high-flow strategy for the collector (maximum temperature rise in the collector about 10°C), are recommended. An example of such a storage tank is shown in Figure 4.19. If the collector areas are larger, external heat exchangers are preferred, because they can exchange higher thermal capacities with small temperature drops. In the case of the internal heat exchanger, the connecting tube to the collector heat exchanger is mounted at the bottom. The height of the inlet from the collector heat exchanger into the tank varies for different applications.

For high-flow configurations, this inlet can be positioned in the lower part of the tank in order to slowly heat up the tank contents from the bottom to the top. A high-flow configuration typically leads to a temperature rise of about 5 - 10°C and resulting flowrates through the collector are in the range of 30 - 70 litres/m².

For low-flow configurations, which lead to a maximum temperature rise in the collector of up to 40°C, the inlet should always be placed at the level with the same temperature as the collector outlet temperature. This can be achieved with special devices, known as stratifiers, as shown in the example of Figure 4.20. Low-flow systems are often applied for domestic hot water preparation or combined solar systems for hot water and space heating, since a large temperature rise is needed to heat up the cold water coming from the mains to the desired hot water temperature.

FLOW DESIGN

Figure 4.20
Stratifying units for hot water storage (Source: Solvis, Germany).

Although advantageous in solar domestic hot water systems, low-flow design is not recommended in solar-assisted air-conditioning systems. The basic reason is that thermally driven cooling equipment in general works at comparatively low temperature differences between inlet and outlet of e.g., 10°C, whereas low-flow regimes produce much larger temperature differences, at least at high radiation levels. Another reason arises from the fact that temperatures up to 100°C are used in solar-assisted air-conditioning. However, temperatures above 100°C are not feasible if the tank is operated at atmospheric pressure, which is likely in order to keep the investment cost low. Under these conditions, the large temperature differences of up to 40°C achieved by low-flow operation are unfavourable. In the worst case, the temperature in the tank would not be raised above 60°C by the collector, because temperatures above 100°C would occur at the collector outlet and the collector pump would be switched off by the control unit in order to prevent boiling.

To avoid this, matched-flow systems are often used in solar-assisted air-conditioning, where the mass flow through the solar collector array is varied such that the collector outlet is about 10°C higher than the required operation temperature of the thermally driven chiller or thermally driven desiccant system (e.g., generator, regeneration heat exchanger). When the whole storage volume has already reached this temperature level, the collector is operated in the high-flow mode and the tank is charged to its maximum allowed temperature (e.g., 95°C).

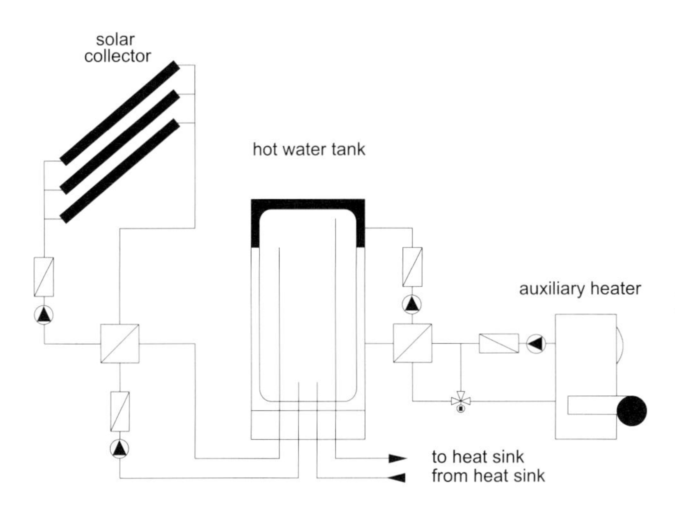

Figure 4.21
Schematic drawing illustrating the integration of a solar collector to a hot water storage with external heat exchanger. Also the connection to the auxiliary heater and to the heat sink is shown.

For matched-flow systems, the inlet from the collector can be positioned close to the top of the tank, as for instance shown in Figure 4.21.

AUXILIARY HEATER LOOP

The inlet pipe from the auxiliary heater should be located at the top of the tank. To determine the outlet position to the auxiliary heater, the following aspects should be taken into account:

- The tank should always contain enough hot water to meet the heat sink demand. The volume for the auxiliary heater can be calculated from the power of the auxiliary heater and the maximum heat sink demand over a specific time period.

- The auxiliary heater often needs a minimum operating time (especially solid wood burners). The volume between the auxiliary heater inlet and outlet must allow for this minimum operating time.
- The outlet to the auxiliary heater should be positioned as high as possible in the storage tank (taking into account the previous aspects) in order to leave the largest possible water volume in the storage tank for the solar collector.
- If the auxiliary heater has enough power, it can also be positioned behind the storage tank in the supply to the heat sink. This construction provides maximum storage capacity for the solar collector.

System Integration

Figure 4.21 illustrates the integration of a collector system with a collector area greater than 15 m^2 using a matched-flow design (collector inlet close to the top of the tank), auxiliary burner and the connection to the heat sink (solar-assisted air-conditioning system).

4.2.2 Storages with phase change materials

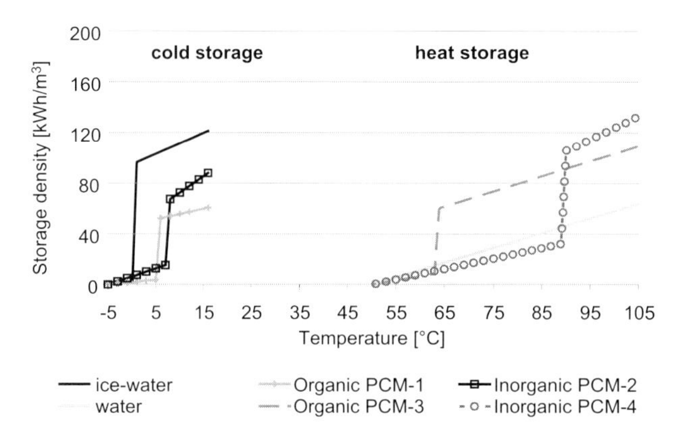

Figure 4.22
Storage density of different phase change materials in comparison to water-ice (cold storage) and pure water (heat storage); only the material related values are shown excluding all equipment associated with storage, such as heat exchangers, insulation etc.. Minimum temperatures of -5°C and 50°C have been taken as references for cold and hot storage, respectively. The following phase change materials were used in the figure /4.7/:

Organic PCM-1:	*Tetradecane*
Inorganic PCM-2:	$Na_2SO_4 \cdot 10 \cdot H_2O/KCl/NH_4Cl$
Organic PCM-3:	*Nonacosane*
Inorganic PCM-4:	$Mg(NO_3) \cdot 6 \cdot H_2O$

The thermal energy absorbed by a material when changing its phase at a constant temperature is called 'latent heat'. For practical applications, materials that exhibit low volume changes are used, for example, solid-to-liquid and some special solid-to-solid phase change materials are applicable. During the phase change, a large amount of energy is absorbed without a temperature change.

Materials

The commonly used phase change materials for technical applications are: paraffins (organic), salt hydrates (inorganic) and fatty acids (organic). For cooling applications, it is also possible to use ice storage; of course, chillers producing cold temperatures below 0°C are necessary in order to apply this kind of a storage strategy. This is not possible using common thermally driven chillers like absorption machines with water-LiBr or adsorption chillers. Phase change materials can be incorporated into a thermal storage system either on the cold or hot side of the thermal cooling unit, depending on the melting temperature of the phase change material.

Figure 4.22 shows materials for storage application in the low temperature range (cold side) on the left hand side and materials for heat storage application (hot side) on the right hand side. It is evident from the data that latent heat storage offers a significant advantage if only a small temperature difference is usable, since in those cases the corresponding storage density of water is small.

Construction Techniques

In all cases, heat must be transferred between the phase change material and the fluid cycle (charging, discharging). To optimise the heat transfer processes, different techniques are used in practical applications, including:

- Direct contact between phase change material and heat transfer fluid: this is possible only if the two materials are chemically stable for long periods of direct contact and the solidification does not occur in a uniform block, preventing sufficient heat transfer during subsequent melting.
- Macroscopic-capsules: this is the most frequently used encapsulation method. The most common approach is to use a plastic module, which is chemically neutral with respect to both the phase change material and the heat transfer fluid. The modules typically have a diameter of some centimetres. An example of a capsule is shown in Figure 4.23.
- Micro-encapsulation: this is a relatively new encapsulation technique in which the phase change material is encapsulated in a small shell of polymer materials with a diameter of some micrometres. A large heat-exchange surface results and the powder-like spheres can be integrated into many construction materials. The technique is only feasible with materials which are not soluble in water, as it is for instance the case for paraffins. However, this technology is still in the development phase. The first commercial products may enter the market in a few years.

Advantages and Disadvantages

The main advantages of phase change storage in comparison to conventional water storage techniques are:

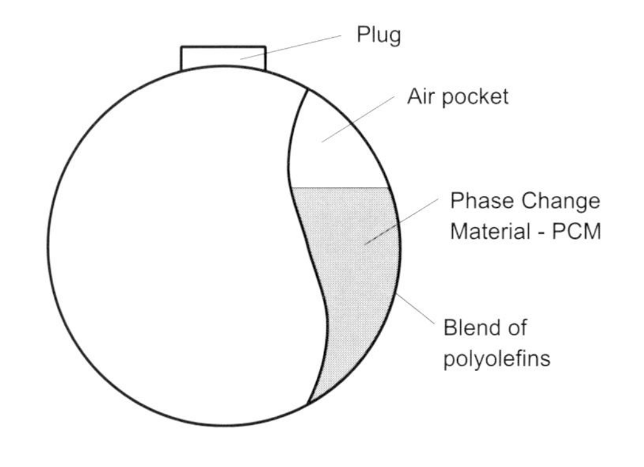

Figure 4.23
Example of a phase change material capsule with a diameter of some centimetres (Source: Cristopia, France).

- Higher thermal energy storage capacity than sensible energy storage, at least if only small useful temperature differences can be achieved; this significantly reduces the volume required for a latent heat storage unit compared to a conventional hot water storage unit.
- Relatively constant temperature during charging and discharging; this may result in a considerable increase of the solar collector efficiency, since the collector performance decreases with increasing operating temperature. Depending on the phase transition temperature also the storage losses may be lower than for water storage because of the more uniform storage temperature.
- Burner cycles for the back-up heat generation unit can be reduced.

The main disadvantages of phase change storage are:

- Higher investment cost, in most cases, compared to water storage.
- In many cases, the peak power during discharge is limited. The reason is the growing thickness of the solid layer, which causes an increasing heat resistance during discharge. This is the main limit determining the acceptable size for the storage modules.
- Limited experience with long-term operation (after many thousand cycles).
- Risks of loss of stability of the solution and deterioration of the encapsulation material.

PART II - SYSTEMS

5 SYSTEM CONFIGURATION - EXAMPLES, CONTROL AND OPERATION

In the previous chapters the components of a solar-assisted air-conditioning system were described. This chapter focuses on the classification of different schemes for solar generation of cooling power. The most important configurations are described and their main operation strategies are outlined. The main focus is on the solar energy supply and its consequences on design, operation and control of the air-conditioning equipment. In many projects, the design constraints for the air-conditioning and solar plant could lead to specific design schemes, which can be derived from the most common configurations outlined in this section. These are not intended to represent all the possible or technically feasible configurations, but to give the reader the feeling of what can be applied in the majority of cases.

For example, the choice might be to design the solar heat production sub-system with water collectors but without a heat storage unit. However, in the vast majority of cases, the presence of the storage unit in such systems is required to improve efficiency. Therefore, in the following characterisation, mainly solar systems (with liquid collectors) including a storage tank are considered. Nevertheless, particularly in applications such as systems coupled with high-inertia distribution systems, the storage tank on the solar side can be relatively small or even be avoided. For example, systems using concrete slabs are characterised by high inertia, and if coupled with a solar-driven chiller system, can work without a solar heat storage unit. An example of a system with a high thermal mass on the load side is the solar cooling plant of a wine cellar in Banyuls (France) /5.1/.

The selection of an appropriate system should be based on different criteria such as performance and capacity requirements, architectural and space constraints, and economics. The selection of the appropriate air-conditioning system configuration for a certain application, i.e., cold distribution-network requirements, compulsory ventilation, exhaust system, will then be the basis for selecting the appropriate solar cooling technology in order to fulfil the criteria mentioned above. Unlike for standard cooling technology, in solar-assisted air-conditioning systems the cooling/ventilation equipment, the solar plant components and the respective back-up heat or cold sources have to be selected in parallel.

5.1 Solar-assisted and solar-thermally autonomous systems

A fundamental decision, which is independent of the air-conditioning technology selection, concerns the fraction of solar energy use relative to the entire plant energy needs. The possible options are a solar-driven system with no other heat source and no other cold production device or a hybrid system consisting of a solar heat source and any type of back-up. The selection of one of these options has a major impact on the design process.

It is important to point out that the 'solar-thermally autonomous' systems refer to systems where the main driving energy source, i.e., the heat for the thermally driven cooling, comes from the sun. In particular, these systems are not meant to be completely self-sufficient, but always need some sort of conventional energy source for their operation, e.g., for the fans or pumps. The completely self-sufficient systems that work using solar energy exclusively, for example, employing photovoltaics to generate the electricity used by the system components and control system, are not considered in this handbook. The reason is that air-conditioning systems are generally installed in areas with an existing energy infra-structure and a well-established electricity grid. Of course, exceptions may occur, for instance hotels on small islands.

Solar-Thermally Autonomous System

Solar air-conditioning plants, employing solar heat as the energy source, are grouped into two main categories. The first category, which is referred to as 'Solar-Thermally Autonomous Air-Conditioning System', has the following characteristics:

- The background of such a concept is ,'Get what you can', i.e., any amount of available solar energy is used for air-conditioning if it is needed. The main goal is to provide improved indoor comfort conditions with maximum use of solar energy.
- This kind of system might not meet the required cooling demand under all circumstances. This means that sometimes the system will not meet the cooling loads and provide the desired indoor conditions, i.e., indoor temperature and humidity. However, this type of system will at least improve comfort conditions compared to no air-conditioning at all.
- The solar part of such systems has to be designed by applying statistical analysis, assessing how often indoor humidity and temperature values will exceed specific comfort requirements for a given load and a given site. This is usually done by performing annual simulations of the entire system consisting of the solar heat source, the cooling equipment and the building with internal loads etc., using the environmental parameters of the given site. Only a combined simulation of the building and its cooling equipment can correctly model the interaction between both and provide a realistic picture of the resulting indoor air conditions. Therefore statistical analysis of the simulation results - i.e., weighting the hours during which certain set values of indoor temperature and/or humidity fail to meet the desired comfort conditions - makes it possible to decide whether this kind of system is acceptable.
- In general, this kind of system is only suitable if the solar gains and cooling loads are fairly well synchronised.

The solar-thermally autonomous systems are usually applied when a back-up is not feasible or advisable. Typical cases are areas where access to other energy sources (other than solar) is for some reasons not easy or economically attractive (e.g., gas on an island) or simply not wanted or not necessary. An example of such a system is the above mentioned wine cooling system /5.1/ and another example is described in Section 8.3 of this handbook.

Solar-Assisted System

The second category, which is referred to as 'Solar-Assisted Air-Conditioning System' has the following characteristics:

- The main goal of this system is to meet the cooling loads (i.e., guarantee certain indoor comfort conditions) and to reduce the conventional energy consumption by introducing solar heat as the major driving energy source for the system.
- Such systems can be able to meet the required cooling load under all circumstances if the back-up system is correctly dimensioned. This means that desired indoor comfort conditions - i.e., indoor temperature and humidity - can be always met or, if they can not, this limitation is not due to the use of solar energy as a heat source.
- The solar part of such systems has to be designed with the aid of energy balance assessment. A comparison with a conventional reference system can provide an estimate of possible energy savings and thus support the decision making process about the solar system design (see Chapters 6 and 7). A common approach is to carry out an annual simulation (e.g., using time steps of one hour) for the system consisting of the solar heat source and the cooling equipment. Since cooling (and heating) loads are always covered, an integrated simulation including the building is not necessary and cooling loads can be independently calculated.

- Conditions with cooling loads that are well correlated in time to available solar gains favour this solution also; however, this requirement is not as strict as for the solar-thermally autonomous air-conditioning system.

BACK-UP SYSTEM

In general for solar-assisted air-conditioning systems, an important decision has to be made about the kind of back-up system: the back-up can be either a second heat source and use the same thermally driven cooling equipment or it can be a second source for generating cooling power, such as a conventional compression chiller. Which of these options is more appropriate depends strongly on the boundary conditions, the dimensioning of the solar system and the related overall energy balance, as will be shown in Chapter 7. The choice among the latter options has to take into account the availability of other heat sources such as conventional burners using hydrocarbon fuels, waste heat or heat from cogeneration plants. Different technical solutions and their implications on system design, control and operation will be discussed in the following sections.

5.2 Characterisation of solar-thermally driven cooling systems

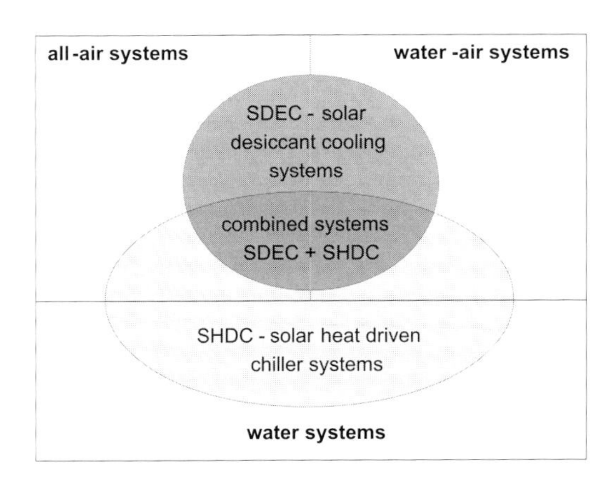

Figure 5.1
Air-conditioning system types and solar-driven technological options.

As already discussed, different types of technology can be applied in air-conditioning systems. The three main technological types, classified according to the cooling medium used in the distribution system and the indoor air-conditioning equipment were presented in Chapter 2. Accordingly, the designer has to select a configuration within the above-mentioned options in order to fulfil the specific application requirements. Based on the selected system configuration, a decision must be made about which solar-operated cooling technology should be used. Solar air-conditioning plants are mainly based on solar heat-driven chillers (SHDC) or solar desiccant and evaporative cooling systems (SDEC). The chillers discussed in Section 3.1 are used to produce chilled water and for this reason can be employed in any of the air-conditioning systems described in Chapter 2, i.e., all-air, water-air and only water systems. The solar desiccant and evaporative cooling systems (SDEC), based on the technology described in Section 3.2, are characterised by the use of desiccant materials, which enhance the direct evaporative cooling potential of the system under given conditions. In relation to the air-conditioning system characterisation given in Chapter 2, these systems could be employed only if a ventilation system is present, i.e., in all-air and water-air systems. In the further text they are called 'solar desiccant cooling systems'.

SYSTEM CLASSIFICATION

Figure 5.1 shows the possible coupling of an air-conditioning system with a particular type of solar-driven air-conditioning technology. A solar desiccant cooling system (SDEC) can be employed in an all-air system or it can be used for primary air treatment in a water-air system. A solar heat driven chiller system (SHDC) can either provide chilled water to the cooling coil in an air-handling unit (all-air-system) or supply chilled water for a water-based air-conditioning system (water system). It also can provide chilled water to a water-air system for the cooling coil in the air handling unit and/or in the room terminals. The overlap area in Figure 5.1, SDEC + SHDC, represents applications where both solar desiccant cooling and solar heat driven chiller technology are jointly employed; these systems are further referred to as combined systems. Generally, all shown systems also allow use of available solar heat for meeting the heating demand.

In the next sections, system configuration examples are shown for each type, including some general suggestions for the corresponding control strategies.

5.2.1 Solar desiccant cooling systems (SDEC)

Once the decision for an air system has been taken, either a conventional air-handling unit or an air-handling unit that employs desiccant components can be used. Applying desiccant cooling technology often results in advantages for the thermodynamic process to bring air to the desired comfort conditions. In fact, in comparison with traditional systems - i.e., employing liquid transport medium and cooling coils - these systems do not need to fall below the dew-point temperature in order to perform the necessary dehumidification, which generally results in increased energy efficiency /5.2/, /5.3/.

Desiccant cooling systems use air as the cooling medium. They are mainly based on the possibility of enhancing the evaporative cooling potential under given conditions by exploiting thermal energy for the regeneration of the dehumidifying component. The system shown in Section 3.2 is the schematic basis for further more complex adjustments that the designer can work out when customising the plant configuration. The latter depends strongly on the distribution of the loads (particularly latent loads), the correlation of the solar radiation and cooling loads, plant technical requirements and design constraints.

Basic Classification

An overview of the different options for solar desiccant cooling systems is given in Table 5.1. Three groups of systems are listed: solar-thermally autonomous, solar-assisted with back-up heating devices and solar-assisted in combination with a back-up component on the 'cold' side. The configurations described in the table are the most common desiccant based solutions for air-conditioning applications with ventilation systems, according to the state of the art of the commercially available technologies. Nevertheless, as previously stated, the system designer could find better solutions according to the specific project characteristics and constraints.

system number		liquid collector	air collector	heat storage	back-up heat	desiccant wheel	thermal chiller	back-up chiller	description	application
5.1.1	a		X			X			solar-thermally autonomous desiccant cooling systems with either solar air or liquid collector	no strict requirements for indoor conditons; 'get what you can' strategy; high correlation between solar gains and load necessary
	b	X		X		X				
5.1.2	a		X		X	X			solar-assisted desiccant cooling systems with either solar-air or liquid collector + back-up heat source	only thermally driven; convenient where low-temp. heat is available; application in temperate climates; no high dehumidification
	b	X		X	X	X				
5.1.3	a		X			X		X	solar-assisted desiccant cooling systems with either solar air or liquid collector + back-up chiller	back-up used for cooling as in common air-handling units; sufficient dehumidification even in warm-humid climates; possible to keep comfort in narrow range
	b	X		X		X		X		

Table 5.1
Common typologies of solar desiccant cooling systems

The systems shown in Table 5.1 can employ either solar air or liquid collectors. The choice between the two systems should take some technical aspects into account. In general, air systems are used for both heating and cooling purposes. In systems using air collectors as a heat source, either the air-handling unit must be appropriately modified or a duct diversion has to be added, allowing different hot air-inlet positions according to the operation mode (heating, cooling).

SOLAR COLLECTOR OPTION

It is important to point out that in general no storage unit is installed along with the solar air collectors in the solar system. This implies that solar air collectors are often employed either in systems where there is a high correlation between solar radiation profiles and cooling loads or in systems where activation of building thermal masses by ventilation air is feasible, like, e.g., in systems using night ventilation. For air-heating purposes, they are also often applied for pre-heating, in order to reduce the fossil fuel consumption of a conventional burner. Water collector systems, by contrast, are generally employed in circuits where a heat storage tank is included. Furthermore the parasitic energy consumption, i.e., mainly fans, should be assessed carefully in the case of solar air collectors since it can vary more significantly according to the plant configuration, than for systems employing liquid collectors.

The solar-assisted systems must employ a back-up, either on the cold (i.e., chiller) or on the hot side. In many cases where low-temperature heat is available (cogeneration plants, industrial process waste heat, etc.), a back-up heat system is employed. In those cases, the solar plant is designed to follow the heat demand peaks not covered by the other heat source.

SYSTEM CONFIGURATIONS

The following sections present some typical designs and applications, providing a system configuration and a short description of each one. Three examples are presented, namely:

1. Solar-thermally autonomous desiccant cooling system with solar air collector integrated as well as ambient-air designs (Example 5.1.1)
2. Solar-assisted desiccant cooling system with a solar liquid collector, heat storage unit and back-up heat source (Example 5.1.2)
3. Solar-assisted desiccant cooling system with a solar air collector (or liquid-based collector) and back-up chiller (chiller as a heat pump and chiller cooled by ambient air) (Example 5.1.3)

For each example, a general control scheme of the solar-assisted desiccant cooling unit is presented.

Example 5.1.1 Solar-thermally autonomous desiccant cooling system with a solar air collector (integrated and ambient-air design)

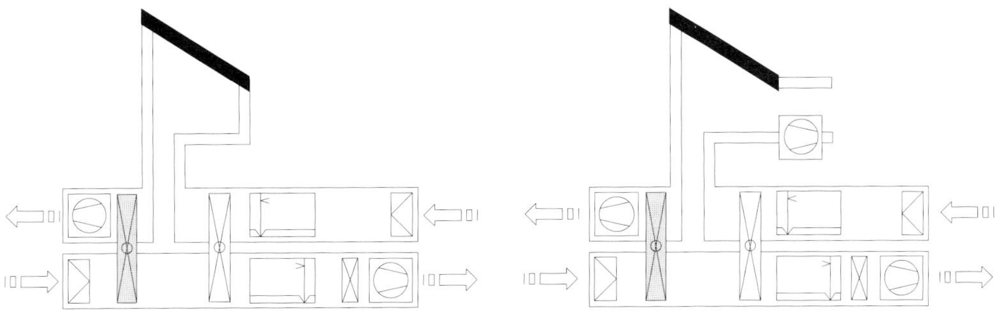

Figure 5.2.a ***Figure 5.2.b***
Solar-thermally autonomous desiccant cooling system with solar air collector, a) integrated design and b) ambient-air design.

DESCRIPTION

This kind of system follows a simple philosophy: as soon as the sun shines, the system can provide conditioned air to the building. However, such desiccant systems can also be used as evaporative coolers only, without employing sorptive dehumidification, allowing a reduction of the supply air temperature if a potential is available, i.e., the environmental air humidity is low enough to make evaporative cooling feasible. Therefore, such a system may be appropriate in temperate climates. Furthermore, the system heat source is the solar air collector field, which implies that the solar energy converted into heat cannot be stored but is used as soon as it is produced. This specific feature of systems that employ solar air collectors makes them suitable for loads that are in phase with solar gains. A typical application might be the all-air air-conditioning system of a seminar room with large glazed surfaces and with a lightweight structure. Such a building has load characteristics which fit well to the plant considered, since both the loads and the energy availability are in phase and because high ventilation air rates are required due to human occupancy /5.4/.

A main decision during the design process concerns the integration of the solar air collector in the plant configuration. There are two possible options (Figure 5.2.a and Figure 5.2.b) with the following specific advantages and disadvantages:

Option 1 integrated design (Figure 5.2.a): the return air from the building is used for regeneration. This means that the relatively high temperature of the return air at the outlet of the heat-recovery unit serves to preheat the regeneration air. However, due to evaporative cooling of the return air before the heat-recovery component, this air also has high humidity, which is less favourable for the regeneration. An advantage of this design is that only one fan serves for both return and regeneration air.

Option 2 design with ambient air for regeneration (Figure 5.2.b): ambient air is used for regeneration. In this concept the temperature increase that can be achieved by the solar collector is higher. However, the temperature level needed for the same regeneration is lower, if the humidity ratio of ambient air is lower than that at the inlet of the solar collector in the 'integrated design'; in general, this is the case in temperate climates. The lower temperature leads to increased solar collector efficiency. A disadvantage is that it is necessary to use another fan for the regeneration air, which results in higher capital and operating costs.

In both cases, as for solar-thermally autonomous systems in general, plant design and configuration choices must be based on annual simulations using climatic data and load analysis.

In general, such a system should be equipped with variable-speed fans since in some cases the supply air conditions will not be adequate to cover the entire cooling loads. The fan speed should be controlled by the indoor air temperature/ humidity levels, using the required air-flow for hygienic needs as a minimum. In addition variable air-flow on the regeneration side will allow regeneration temperature control.

Control / Operation

The operation modes of the considered air-conditioning system implement the physical processes needed to treat the ventilation air according to the building loads. In this way, the system operation is characterised by the active components in the mode, keeping temperature and humidity control together. During the operation modes which implement the heating/cooling and humidification/dehumidification functions, the physical processes shown in Section 3.2 take place in the air-handling unit.

The system operates in following modes:

- Free ventilation mode: none of the thermal components is active; no driving heat is required
- Indirect evaporative cooling mode: The return air stream humidifier is active as well as the heat recovery unit. The return air is brought close to saturation and then enters the heat exchanger. Only sensible cooling of the supply air stream is provided. No driving heat is required. Main control parameters: efficiency of return air humidifier (0 - 100%)

CONTROL / OPERATION

- Combined evaporative cooling mode: the supply air and the return air humidifiers are active. The heat recovery unit is in operation. Combined evaporative cooling, i.e., direct and indirect, is employed. No driving heat is required. Main control parameters: efficiency of supply air humidifier (0 - 100%)
- Desiccant cooling mode: the dehumidifier wheel, the humidifiers, the heat recovery unit and the solar air collector are active; all heat available from the solar system is used for regeneration. Main control parameters: regeneration air temperature by means of control of the fan rotational speed, supply air humidifier (0 - 100%)

Table 5.2 describes the operation scheme. The bypass column included in the table is valid only in the integrated design, since no bypass is needed if there is a decoupling of the streams coming from the air collector through the regeneration part of the wheel and the air coming from the building. In this case, the air flowrate of the two streams can be controlled by the two different fans. In the case of ambient air used for regeneration, it could make sense from a technical point of view to use a bypass instead, and divert part of the stream, which would go through the collector, directly towards the desiccant wheel. This would allow control of the regeneration temperature, and a reduction of the fan electricity consumption due to the decreased pressure drop.

Mode (increasing cooling load ↓)	Components active (+), not active (-)									Condition
	Desiccant rotor	Heat recovery unit	Humidifier supply air	Fan supply air	Humidifier return air	Bypass regeneration air heater	Regeneration air heater	Fan return air	Fan regeneration air (only ambient-air design)	
Free ventilation	-	-	-	+	-	open	-	+	-	supply air temperature and humidity o.k.
Indirect evaporative cooling	-	+	-	+	+	open	-	+	-	supply air temperature exceeds set value
Combined evaporative cooling	-	+	+	+	+	open	-	+	-	supply air temperature exceeds set value; supply air humidity below set-point
Desiccant cooling	+	+	+	+	+	≤20%	+	+	+	supply air temperature and/or humidity exceeds set value; solar heat available

Table 5.2
Operation scheme of a desiccant cooling unit driven with heat coming from a solar air collector (solar-thermally autonomous system)

Example 5.1.2 **Solar-assisted desiccant cooling system including a collector with liquid heat-transfer medium, storage tank and back-up heat source**

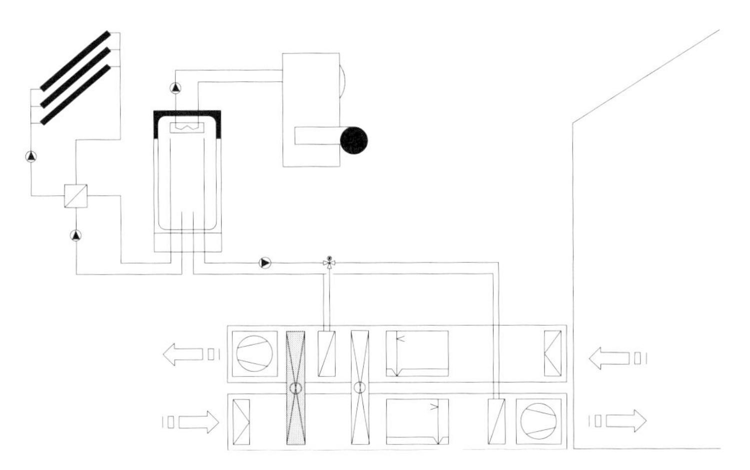

Figure 5.3
Solar-assisted desiccant cooling system including a collector with liquid heat-transfer medium, storage tank and back-up heat source

Description

This is a very common design for a solar-assisted air-conditioning system using the desiccant cooling technique. Solar heat is supplied either to the heat storage tank or directly to the load, depending on its integration in the plant scheme. If the solar heat available is not sufficient, the back-up heat source is used. In the case that the back-up heat source is also connected to the storage tank, the top storage temperature is the parameter used for control purposes, i.e., as soon as this temperature falls below the set value, the back-up heater turns on. If the back-up heater is directly integrated into the water cycle which provides heat to the regeneration air heater, it is switched on as soon as the desired regeneration temperature is higher than the one of the hot water stored. In this design, the full tank volume is available to store solar heat.

In general, such a system should be equipped with variable-speed fans since in some cases the minimum air-flow, i.e., the required flow of fresh air, could not cover the cooling loads. Nevertheless, if the system is designed to cover only part of the loads (i.e., other air-conditioning systems are installed), the air-handling unit could be equipped with fixed-speed fans.

Control / Operation

The general structure of the control scheme is the same as the one described in the previous example; the modes described in the previous section are valid also in this case, as can be seen in Table 5.3.

When the configuration discussed above, i.e., back-up heater integrated into the plant scheme, works in the desiccant cooling mode, it is recommended to adjust the applicable regeneration temperature. In this way the water temperature enters the regeneration air heater according to the actual indoor cooling needs. Using a variable regeneration temperature complicates the control procedure but allows a higher plant efficiency under part load conditions. Moreover, it has two positive effects: the storage tank volume is used better and the solar collectors operate with higher efficiency.

If variable-speed fans are installed, their speed should be controlled by the indoor temperature/humidity unless hygienic needs make a higher volume flow necessary. This means that the control strategy would be based on the one shown in the previous section, but if comfort conditions were not achieved, the air flowrate in the 'desiccant cooling mode' would be increased.

(increasing cooling load ↓)

Mode	Desiccant rotor	Heat recovery unit	Humidifier supply air	Fan supply air (*)	Humidifier return air	Bypass regeneration air heater and desiccant wheel	Regeneration air heater	Fan return air(*)	Condition
Free ventilation	-	-	-	+	-	open	-	+	supply air temperature and humidity o.k.
Indirect evaporative cooling	-	+	-	+	+	open	-	+	supply air temperature exceeds set value
Combined evaporative cooling	-	+	+	+	+	open	-	+	supply air temperature exceeds set value; supply air humidity below set-point
Desiccant cooling	+	+	+	+	+	≤ 20%	+	+	supply air temperature and/or humidity exceed set value
Desiccant cooling increased air-flow	+	+	+	++	+	≤20%	+	++	supply air temperature and/or humidity exceed set value

(*) the sign ++ applies for increased air-flow according to control strategy

Table 5.3
Operation scheme of a solar-assisted desiccant cooling system including a collector with liquid heat-transfer medium, storage tank and back-up heat source

Example 5.1.3 Solar-assisted desiccant cooling system including a collector with liquid heat-transfer medium and back-up chiller (chiller as an integrated heat pump and chiller cooled by ambient air)

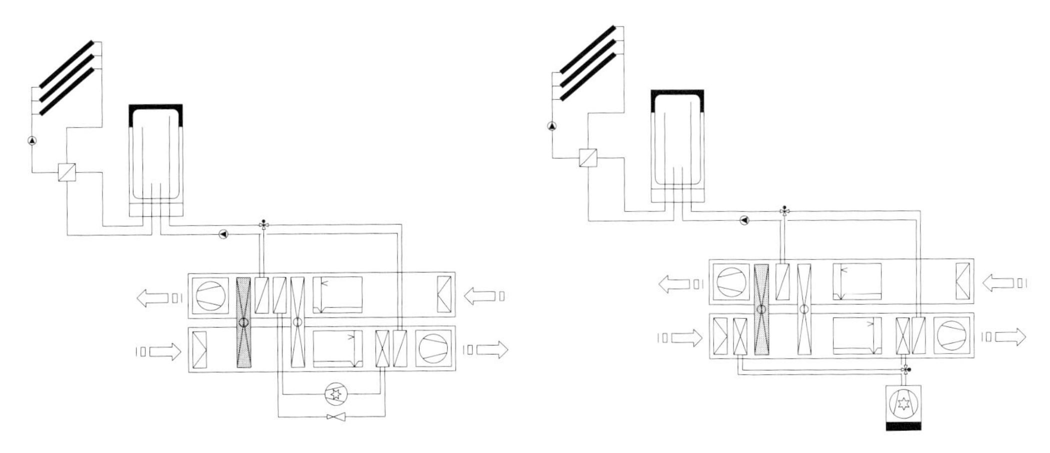

Figure 5.4.a ***Figure 5.4.b***
Solar-assisted desiccant cooling system including a collector with liquid heat-transfer medium and back-up chiller. Two configurations are shown:
a) chiller as an integrated heat pump and
b) chiller cooled by ambient air

SYSTEM WITH HEAT-PUMP

The two systems described here use conventional compression chiller technology as the back-up. In the first case (see Figure 5.4.a), the compression machine is used as a heat pump between the supply and the return air streams. It operates by lowering the temperature of the supply air and delivering the condensation heat to the regeneration air. Therefore a direct evaporator and direct condenser without additional water circuits are used. The advantage of this system is the high heat-recovery rate that can be achieved since the heat pump provides both cooling of the supply air and heating of the regeneration air. The heat pump has to work at a higher compression rate due to the higher temperature difference compared to a machine using ambient air for condensation. Although the supply-air cooler can always provide cooling, it is necessary also to install a humidifier on the supply air side; if enough solar radiation is available the latter allows the plant to be operated as a conventional solar desiccant cooling system as in previous examples.

System With Two Cooling Coils

The other system (see Figure 5.4.b) is a SDEC with two cooling coils integrated into the supply side. The cooling coils are connected to a circuit where a conventional vapour compression chiller is included. It should be noted that another configuration derived from the latter is mostly employed, if the climate conditions are not extreme. In fact, the most commonly used solar-assisted desiccant cooling systems with a back-up chiller include only the cooling coil after the heat recovery wheel. In this configuration, the desiccant component is intended to carry out all of the dehumidification. The chiller will then cover only a part of the sensible load, i.e., the part not covered by combined evaporative cooling.

The configuration with two cooling coils (Figure 5.4.a), is designed for humid climates such as e.g., in tropical areas. Under this kind of weather conditions with high humidity ratios of the ambient air (greater than 20 g/kg), the sorption process is not able to reduce the air humidity sufficiently to achieve comfortable supply-air conditions. Therefore, another dehumidification device is necessary. The system shown here has the advantage that the first cooling coil (coil 1), which is installed before the dehumidifier and which is used for cooling and dehumidification, works at a higher temperature level compared to the case where all the dehumidification is achieved by the chiller. The pre-dehumidified air is further dehumidified in the sorption wheel to the desired level for the supply-air humidity. In this configuration the humidifier for the supply air is not operated when the two cooling coils are used. Since the second cooling coil (coil 2) is only used to cover sensible loads, it can also work at a higher temperature level. This means that the COP of the chiller, which serves both cooling coils, will be higher than in a case where the air-conditioning load is covered entirely by the chiller. Therefore, this kind of system offers more than just desiccant cooling. It is a highly efficient concept in warm and humid climates, to combine solar-driven dehumidification with a conventional vapour-compression chiller.

Mode (increasing cooling load ↓)	Components active (+), not active (-)											Condition
	Cooling coil 1 (cooling + dehumidification)	Desiccant rotor	Heat recovery unit	cooling coil 2 (only sensible cooling)	Humidifier supply air	Ventilator supply air	Humidifier return air	Bypass regeneration air heater and desiccant wheel	Regeneration air heater	Ventilator return air	Back-up chiller	
Free ventilation	-	-	-	-	-	+	-	open	-	+	-	supply air temperature and humidity o.k.
Indirect evaporative cooling	-	-	+	-	-	+	+	open	-	+	-	supply air temperature exceeds set value
Combined evaporative cooling	-	-	+	-	+	+	+	open	-	+	-	supply air temperature exceeds set value; supply air humidity below set-point
Desiccant cooling without chiller	-	+	+	-	+	+	+	≤20%	+	+	-	supply air temperature and/or humidity exceed set value
Desiccant cooling with coil 1 active	+	+	+	-	-	+	+	≤20%	+	+	+	supply air humidity exceeds set value
Desiccant cooling with coil 2 active	-	+	+	+	+	+	+	≤20%	+	+	+	supply air temperature exceeds set value
Desiccant cooling with coil 1 and 2 active	+	+	+	+	-	+	+	≤20%	+	+	+	supply air temperature and humidity exceed set value

Table 5.4

Operation scheme of a desiccant cooling unit driven with heat coming from solar collector with liquid heat-transfer medium and a compression chiller that provides chilled water for two cooling coils (according to the system in Figure 5.4.b).

CONTROL / OPERATION

The general structure of the operation strategy of these systems is the same as for the previous SDEC example. However, the humidifier for the supply-air stream must be controlled in a different way. It works with increasing loads up to the desiccant cooling mode. As soon as further dehumidification is required, it is not operated any more. The latter is valid both for the systems with heat pump and with two cooling coils. Table 5.4 describes the operation scheme of the system employing two cooling coils.

5.2.2 Solar heat driven, chiller-based systems (SHDC)

In Section 3.1, a technical description of sorption chiller components based on closed thermodynamic cycles was presented. In this section, solar thermally driven, chiller-based systems are discussed. All systems in this category use a thermally driven sorption chiller as a main component.

With solar-driven adsorption or absorption chillers, the water is chilled in a closed system cycle. The chilled water can be used in all-air systems, e.g., in an air-handling unit, in water systems, e.g., using chilled ceilings, or in any air-water system, e.g., induction systems (see Chapter 2).

When a chiller-based system is designed for both sensible cooling and dehumidification, for example in combination with fan-coils, typically a chilled-water temperature of 6 - 9°C is necessary, since dehumidification is achieved by cooling the ambient/incoming air below the dew-point. In moderate climates with low humidity, it is also possible to use a system that only provides sensible cooling, since no dehumidification is necessary. An example is a chilled ceiling which is supplied with chilled water of about 15 - 18°C.

Since absorption and adsorption chillers operate with hot water, only collectors with a liquid heat-transfer medium can be used. In general, the system configuration is independent of the collector type (i.e., flat-plate, evacuated tube etc.). The solar collectors should be selected based on the required chiller driving temperature - i.e., hot water supply temperature - and economics. Appropriate design of the collector area is discussed in Chapter 6.

An overview of possible different options for solar-thermally driven air-conditioning with chiller-based systems is given in Table 5.5. As in the previous section, these examples are meant to be used as a guide for most common cases, and are not aiming to cover all the possible configurations. They show configurations which are reasonable from a technical point of view, given the current technology state of the art.

SYSTEM CONFIGURATION

The following sections present some typical designs and applications, providing for each system configuration a short description and a table which describes the different operation and control conditions. Three examples are presented, namely:

1. Solar-thermally autonomous system with thermally driven chiller for chilled-water production e.g., for a chilled ceiling.
2. Solar-assisted system with thermally driven chiller and back-up heat source.
3. Solar-assisted system with thermally driven chiller and electrically driven compression chiller as the back-up.

In SHDC systems, the selection of back-up technology is the same as with SDEC systems, depending on project constraints and boundary conditions. However, it should be noted that generally the primary energy consumption for a thermal chiller is higher than that for a compression chiller (see Chapter 2 and particularly Chapter 7). This, in most cases, makes it more efficient to use a cold back-up than a heat back-up. The previous statement can be considered valid on the basis of SHDC technology commonly employed and the average efficiency values for electricity production in most countries /5.5/.

system number		liquid collector	air collector	heat storage	back-up heat	desiccant wheel	thermal chiller	back-up chiller	cold storage	description	application /example
5.2.1	a	X		X			X			solar-thermally autonomous system; solar thermally driven chiller	high correlation between solar gains and load or lower requirements on indoor conditions; e.g. solar-driven chilled-ceiling system with natural ventilation
	b	X		X			X		(X)		
5.2.2	a	X		X	X		X			solar-assisted system employing a thermally driven chiller with solar collector and back-up heat source	small to medium cooling capacity; the necessity of the cold storage unit depends on the dynamics of the load profile; most commonly employed system in practical application
	b	X		X	X		X		X		
	c	X			X		X				
5.2.3	a	X		X			X	X	X	solar-assisted system using a thermally driven chiller with solar collectors and electrically driven back-up chiller	medium to high cooling capacity; energy and peak saving; e.g. solar contribution to a chilled-water network during peak load periods
	b	X		X			X	X			

Table 5.5
Classification of solar-thermally driven, chiller-based air-conditioning systems

Basic Control

The control strategies of these systems usually comprise four steps which are described briefly here:

1. The primary-circuit pump of the solar system, i.e., the pump which moves the fluid through the solar collector, is switched on when a given temperature difference between collector outlet and tank is reached or the solar irradiation exceeds a specific set-point.
2. When the collector outlet temperature exceeds the storage tank temperature, the tank load pump, i.e., the secondary pump is switched on and the tank is loaded until a minimum driving temperature is reached. If at the same time there is already a cooling demand, the sorption chiller may be activated using an auxiliary heat source or, if present, the back-up chiller would be used for conditioning the supply air.
3. After the minimum driving temperature is reached, the solar driven cooling process can be operated. Depending on the system and the control strategy the sorption chiller may be on or off before this state and it may be kept off if the minimum temperature has been reached but no cooling demand is present.
4. The back-up system operates if the driving temperature is above the minimum temperature but not high enough to supply the total heat demand of the sorption chiller for the cooling capacity, which is required to meet the load. If there is no back-up system, then two strategies are possible:
    a. Keeping the chilled-water temperature constant at a reduced cooling capacity or
    b. Maximising the cooling capacity at variable chilled-water temperatures (see the example of a solar-thermally autonomous system for more information).

It is good practice to use hot and cold storage unit temperatures as indicators to switch between the different stages. Beside increasing the stability of the control system, storage units can improve the total system performance mainly by levelling load shifts.

Heat Storage

Because of the sorption chiller's inertia and the thermodynamically unavoidable losses it is advisable to install a hot water tank in the solar circuit. To achieve a continuous and thereby effective cold generation process even under variable solar irradiation, a minimum tank volume is required.

On the other hand, the storage volume should not be overdimensioned because a minimum storage temperature has to be reached before the chiller can be started. If this preheating phase takes too long, like e.g., in early morning hours, the advantage resulting from the synchronisation of the cooling load and solar radiation may be lost. This may have a great impact on solar-thermally autonomous systems. In solar-assisted systems, it may also result in significant increase of the back-up energy use, if cold generation by the solar system is only possible when the cooling load is already decreasing, because it took very long to heat up the storage to the minimum driving temperature.

COLD STORAGE

The advantage of a cold storage unit, especially when compression chillers are used, is illustrated in Example 5.2.3.

HEAT REJECTION

All the SHDC systems include a cooling water system for heat rejection, which is normally connected to a cooling tower. This circuit is intended to keep the cooling water within a certain temperature band according to the specific technology employed. This is actually the third heat source (represented by the external environment), which makes operation of such technology possible. In water/lithium bromide absorption chillers, it is important to avoid rapid changes in the cooling water inlet temperature in order to prevent crystallisation. This is done by the cooling water valve, which is controlled to an appropriately set temperature (normally between 25°C and 35°C with a deviation of approximately 1°C). The cooling water valve is not essential for all types of chillers, for example adsorption chillers, since no crystallisation can occur.

Example 5.2.1 Solar-thermally autonomous system with thermally driven chiller for chilled-water production

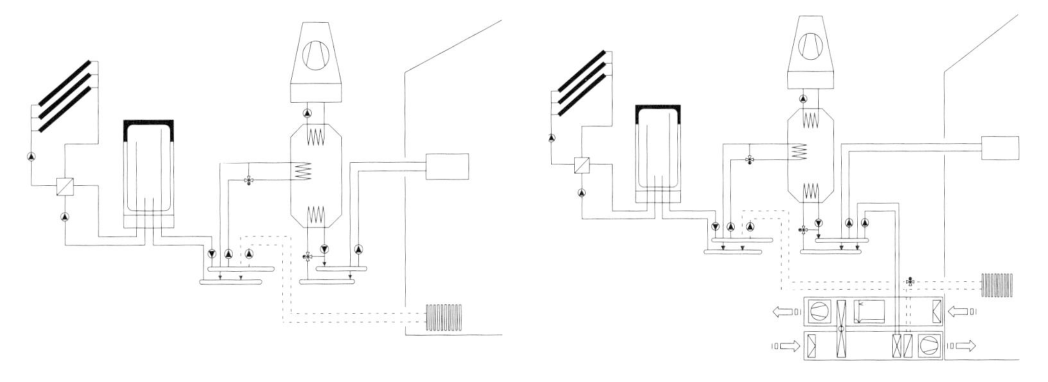

Figure 5.5.a ***Figure 5.5.b***

Solar-thermally autonomous systems with thermally driven chiller for chilled-water production,
a) with chilled-water distribution network and
b) with an air-water system (chilled-water distribution network and air-handling unit).

DESCRIPTION

This kind of system configuration can be applied either in conjunction with a chilled-water distribution network (Figure 5.5.a), or with an air-water system consisting of a chilled-water distribution network and air-handling unit, (Figure 5.5.b). The following discussion is valid for both options.

An important advantage of solar-thermally autonomous systems is that they always guarantee primary energy saving compared to conventional cooling systems (see Chapter 7). In addition, they need less design and installation effort since the configurations are usually simpler. However, care has to be exercised during the design process, because the different inertia of the sub-systems (solar heat production, cooling facility and building) may result in complex behaviour of the entire system. In addition, they may increase the mismatch between the solar-generated cold production and the load.

DESIGN

For this reason, as mentioned in the previous section, annual simulations of the entire system should support design activities. This will allow statistical analysis of the data, which will drive the final design decisions. A main decision during the design process concerns the supply water temperature needed, i.e., the temperature of chilled water. The cooling capacity of a sorption chiller raises with increasing chilled-water temperature at constant driving and cooling water temperatures. Therefore, it is advisable to design a system with a high chilled-water temperature set-point. On the other hand, if dehumidification is necessary, the chiller set-point temperature is chosen as a function of the environmental and supply air conditions, which determine whether or not the air has to be cooled below the dew point.

Since solar-thermally autonomous systems follow the strategy 'Get what you can', there is the possibility that the plant will not be able to cover all the cooling loads at a certain chilled-water temperature, under all conditions; therefore a decision has to be taken whether the capacity can be maximised or the chilled-water temperature has to be controlled. This decision is the main point determining the control strategy of such systems.

CONTROL/ OPERATION

As already discussed, there are two main operation strategies, depending on the comfort target. In both cases, the valve on the hot water side, used to regulate the driving temperature of the chiller, is controlled in order to keep the evaporator outlet temperature within the set band, e.g., by means of a dew-point sensor to avoid condensation in rooms with chilled ceilings. In cases when the heating capacity is not enough to cover the loads completely, this valve will be fixed in the completely open position, in order to utilise the maximum possible driving heat in the generator.

The cooling system connected to the cooling tower is operated as soon as the sorption chiller is turned on.

	Mode	Components active (+), not active (-)								Condition
		Collector pump	Hot storage temperature higher than min. driving temperature	Thermal back-up	Sorption chiller	Hot water side valve	Chilled water side valve	Compression chiller	Load (approx.)	
increasing cooling load ↓	Warm up solar circuit	+	-	-	-	-	-	-	0%	room or supply air temperature and humidity o.k.; cold water is not needed
	Compensate missing capacity in the morning	+	-	-	-	-	-	-	0 - 10%	cold water is needed but temperature of solar hot water is still too low
	Solar cooling	+	+	-	+	+	100%	-	0 - 100%	solar hot water temperature is high enough and cold water temperature is low enough to cover the cooling load
	Compensate missing capacity during peak load	+	+	-	+	100%	adjusted	-	60 - 100%	solar hot water temperature is not high enough or cold water temperature is not low enough to cover the cooling load

Table 5.6.a

Operation scheme of a solar-thermally autonomous system with SHDC: constant chilled-water temperature.

Mode	Collector pump	Hot storage temperature higher than min. driving temperature	Thermal back-up	Sorption chiller	Hot water side valve	Chilled water side valve	Compression chiller	Load (approx.)	Condition
Warm up solar circuit	+	-	-	-	-	-	-	0%	room or supply air temperature and humidity o.k.; cold water is not needed
Compensate missing capacity in the morning	+	-	-	-	-	-	-	0 - 10%	cold water is needed but temperature of solar hot water is still too low
Solar cooling	+	+	-	+	+	100%	-	0 - 100%	solar hot water temperature is high enough and cold water temperature is low enough to cover the cooling load
Compensate missing capacity during peak load	+	+	-	+	100%	100%	-	60 - 100%	solar hot water temperature is not high enough or cold water temperature is not low enough to cover the cooling load

(Components active (+), not active (-); rows ordered by increasing cooling load ↓)

Table 5.6.b

Operation scheme of a solar-thermally autonomous system with SHDC: maximised cold production

Two control strategies are available:

CONSTANT CHILLED-WATER TEMPERATURE

i) Keeping the chilled-water temperature constant at a reduced cooling capacity (Table 5.6a): if the solar driving heat temperature at the top of the hot water tank is not sufficient to cool down the entire volume of water flowing through the evaporator, i.e., from the return temperature to the chilled-water set-point, a valve can be used to bypass the load. The valve is used to mix the evaporator outlet and return flows, which causes a reduction of the evaporator inlet temperature compared to the return temperature. In this way the limited cooling capacity is enough to reduce the (lower) temperature and the supply water matches the set value. Due to the smaller net volume flow to the chilled-water distribution sub-system, there is part of the load which cannot be covered because of the lack of cooling capacity. During normal operation, where the cooling capacity is high enough to meet the cooling loads, the valve on the chilled-water side remains in its fully open position and the chilled-water temperature is controlled by the hot water valve.

This strategy is applied when the removal of latent heat is more important than that of sensible heat.

MAXIMISED COOLING OUTPUT

ii) Maximising the cooling capacity at variable chilled-water temperatures (Table 5.6b): if there is no strict temperature requirement, e.g., for dehumidification, control of the chilled-water temperature by means of the chilled-water valve is not necessary. The latter is maintained in the fully open position. During the preheating phase of the generator and during peak load periods, the chilled-water outlet from the evaporator may be above the temperature needed to meet the load. If a chilled-water valve is installed nevertheless, it should be open and fixed at its fully open position. Even if the cold demand can not be entirely covered by this strategy, the mean evaporator temperature is higher than in case (i) and the reduction of cooling capacity is therefore smaller. This strategy is suitable, e.g., for chilled ceilings, chilled floors or walls and other silent cooling systems.

Example 5.2.2 **Solar-assisted cooling using a thermally driven chiller heated by a solar collector (liquid heat-transfer medium), back-up gas heater and hot water tank.**

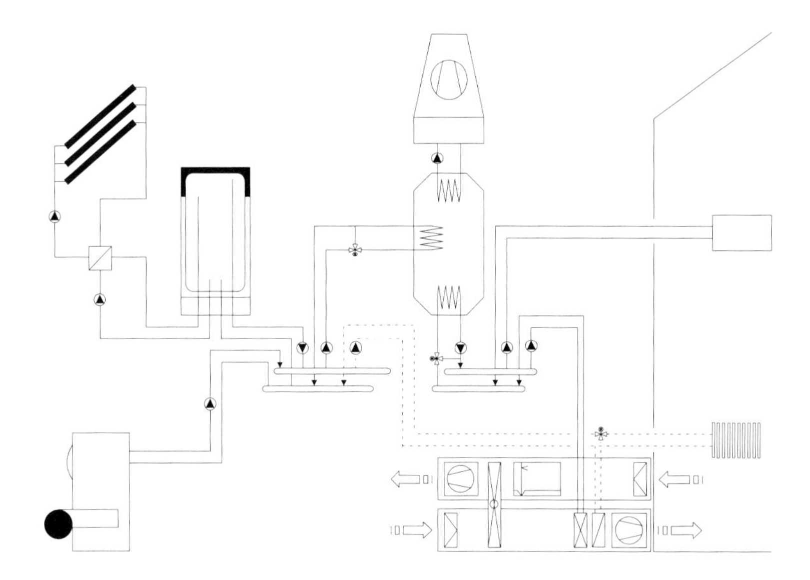

Figure 5.6
Solar-assisted cooling using a thermally driven chiller heated by a solar collector (liquid heat-transfer medium), back-up gas heater and hot water tank.

DESCRIPTION

This configuration consists of a thermally driven chiller connected to a chilled-water distribution network and to an air-handling unit that is used for heat recovery and supply of conditioned fresh air. In the latter the chilled water is used in a cooling coil. The driving heat for the sorption chiller comes from the solar heat production sub-system, which includes the collector, the tank and the back-up heater.

Although this system is the most frequently used configuration among SHDC, it is not easy to give a general description. This is because the kind of chiller used (absorption or adsorption) and the chilled-water demand profile with the specific required chilled-water temperature values have a major impact on the performance and consequently on the system configuration details.

In these systems, dehumidification is achieved by cooling the air stream below the dew-point temperature. As a consequence, in cases where a high dehumidification rate is required and the humidity set-points correspond to low dew-point temperatures, the dehumidified air cannot be blown into the building without further re-heating. A heater has to be installed to re-heat the air to the desired supply air temperature, resulting in an increase of the energy consumption. Even if the re-heating could be done by means of heat from the heat transfer medium at the chiller generator outlet, which leads to a lower collector inlet temperature and then to better collector efficiency, the thermal energy needed to generate the desired supply air conditions would be higher. In principle, also part of the waste heat that is rejected at the cooling tower could be used for re-heating.

BACK-UP HEATER CONNECTION

In general, the back-up heat source can be installed in different ways. Either it delivers heat to the heat storage unit, as for instance shown in Figure 5.3 or it delivers heat directly to the load, as shown in Figure 5.6. In the first case only one connection between heat sources and heat load (chiller generator) occurs, while in the latter case, two heat sources work parallel, namely the solar-storage sub-system and the back-up heat source, e.g., a gas burner. In the first case, the solar collector can work as a pre-heater, i.e. raising the return flow temperature from the heat load to an intermediate level, while the back-up heat source raises it to the final temperature level that is actually required by the control system. In the second case, the solar-storage sub-system can only provide heat, if the actual requested temperature level is achieved. As long as this is not the case, the back-up heat source is switched on.

Each concept has some advantages and disadvantages. In the first case the temperature level to be achieved by the solar system is somewhat lower. However, since the back-up heater also uses the storage tank, the available storage volume for the solar system is reduced.

In any case for both concepts, it is recommended to install a heat storage unit between the load and the solar system if an adsorption chiller is used, in order to level out the peaks in the return flow temperature, which generally occur during the chiller interchanging phase (see Figure 3.7). If storage is not integrated here, the peaks would cause control problems for both the solar collector and the back-up heating system.

The COP of sorption chillers increases as the desired chilled-water temperature increases. This is for example the case, when a chilled-ceiling system is employed, which operates with chilled-water supply temperature of about 16°C. The air-flow needed (which has to be cooled and re-heated) using a chilled ceiling can be reduced in comparison to other cold distribution technologies. Nevertheless, it is interesting to note that in systems such as the one considered here, the higher chilled-water temperature requirement in the chilled ceiling does not bring any advantage. In fact, the efficiency of the sorption chiller is still restricted, since the evaporator outlet temperature has to be approximately 6 - 9°C in order to serve the cooling coil in the air-handling unit. Therefore the system performance is higher, i.e., a higher achievable COP is possible, in situations where the dehumidification demand is not covered by the SHDC system. The latter applies for instance for combined systems (SHDC + SDEC, see Example 5.3.1). If the plant is designed in order to cover dehumidification loads as well, the system control should allow variable chilled-water temperatures so that the air-handling unit cooling coil can also operate at a higher temperature. This would result in significant energy savings under conditions where no dehumidification is necessary.

CONTROL/ OPERATION

Due to the presence of the thermal back-up system, the chilled-water temperature can always be controlled by modulation of the chiller driving temperature. Therefore, a valve on the chilled-water side can be omitted. Nevertheless, during the pre-heating phase, the chiller evaporator outlet temperature is usually above the set value. If this has to be avoided, a chilled-water valve is necessary. Through the valve, some of the chilled water is returned to the evaporator inlet until the outlet temperature matches the set value. After the pre-heating phase (in normal operation mode), the valve is set to the fully open position. In Table 5.7, the chilled-water valve is therefore shown in brackets.

In order to maintain high heat transfer coefficients in the generator, many manufacturers recommend keeping the hot-water flowrate constant. Under this condition, the hot water valve is the main controlling device of the sorption chiller. It is active at any time that the chiller is active and controls the evaporator outlet temperature by adjusting the generator inlet temperature. In some cases, a variable hot water flow can be used, which may lead to lower return temperatures.

Mode	Components active (+), not active (-)								Condition
	Collector pump	Hot storage temperature higher than min. driving temperature	Thermal back-up	Sorption chiller	Hot water side valve	Chilled water side valve	Compression chiller	Load (approx.)	
Warm up solar circuit	+	-	-	-	-	(-)	-	0%	room or supply air temperature and humidity o.k.; cold water is not needed
Compensate missing capacity in the morning	+	-	+	+	+	(100%)	-	0 - 10%	cold water is needed but temperature of solar hot water is still too low
Solar cooling	+	+	-	+	+	(100%)	-	0 - 100%	solar hot water temperature is high enough and cold water temperature is low enough to cover the cooling load
Compensate missing capacity during peak load	+	+	+	+	+	(100%)	-	60 - 100%	solar hot water temperature is not high enough or cold water temperature is not low enough to cover the cooling load

increasing cooling load ↓

Table 5.7
Operation scheme of a thermally driven chiller with solar collectors and a back-up heat source

Example 5.2.3 Solar-assisted system with thermally driven chiller and an electrically driven chiller as a back-up

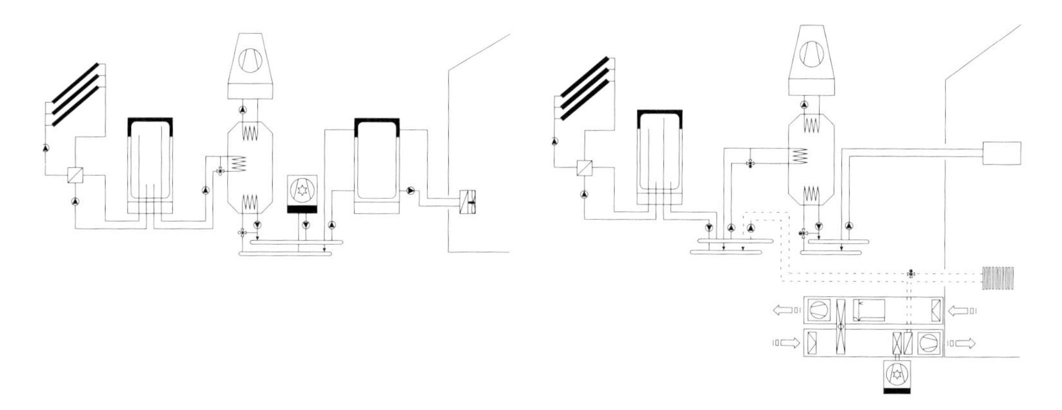

Figure 5.7.a ***Figure 5.7.b***
Solar-assisted system employing a thermally driven chiller and an electrically driven chiller as back-up,
a) with chilled water storage and a fan-coil system
b) with an air-handling unit and a chilled ceiling

Description

Two possible configurations of SHDC systems employing a conventional chiller as a back-up are described in this example. The first configuration includes a chilled-water storage unit and an electric vapour compression chiller, which is installed in parallel to the SHDC and used to cool down the chilled-water storage unit when there is not enough solar heat. The system is then connected to a distribution network with fan-coils. In this case, the fan-coils will cover latent and sensible loads. Given the characteristics of the technology used for distribution (low inertia), and the need to carry out dehumidification, which implies clear control of the chilled-water temperature, installation of a chilled-water storage unit is strongly recommended. The latter is supplied by both chillers and is connected as a hydraulic switch to the distribution network.

In order to avoid that the plant would always operate under unfavourable conditions, i.e., with an evaporator temperature between 6° C and 9° C, variable chilled-water set-point temperatures should be allowed (see previous example). This requires respective design of the fan-coils, since they have to be able to operate at higher temperatures if no dehumidification is needed.

The second configuration shows a system where a SHDC system is coupled with a chilled ceiling network. In addition, an air-handling unit is included in this configuration. The vapour compression chiller is connected independently from the thermally driven chiller to the cooling coil of the air-handling unit. In this scheme, the ambient-air flowrate can be reduced to the minimum required rate and the remaining sensible cooling load is removed by the chilled ceiling.

This scheme has the advantage that the SHDC can provide chilled water at a higher temperature, since the supply air is dehumidified in the air-handling unit. Therefore, a higher COP is achieved than in the previous case. Since the inertial chilled ceiling is supplied directly by the SHDC a chilled water storage is not necessary. In chilled ceilings, the inlet water temperature can vary within a relatively wide band, so a chilled-water valve may be omitted.

CONTROL/ OPERATION

The configuration with chilled-water storage demands a more complex control strategy. The main control parameter is the chilled-water storage temperature, which has to match a load-dependent set-point (above or below the dew-point). A chilled-water valve is necessary for the sorption chiller, since the chilled-water outlet temperature has to be controlled, when the solar heat capacity is not high enough to cool down the entire water flow to the set-point (e.g., below the dew-point). The valve therefore reduces the net flowrate to the consumer until the set-point is reached. Nevertheless, at cooling loads higher than the solar cooling capacity, the storage temperature increases and the compression chiller is switched on. If the system allows variable cold water temperatures, a cold water valve for the compression chiller may be necessary as well. Care has to be exercised when the compression chiller is switched on before the sorption chiller is operating. This is typically possible during the morning hours, when reaching the minimum driving temperature is delayed compared to the demand. Even when the pre-heating phase has started, i.e., generator and chilled-water pump have been switched on, the compression chiller should remain in operation until the chilled-water outlet temperature matches the set-point over a specified time period. It is advantageous if the compression chiller capacity can be reduced step-by-step. If this is not possible, the cold storage volume must be dimensioned appropriately to avoid high-frequency cyclic operation of the sorption and compression chiller.

Mode (increasing cooling load ↓)	Components active (+), not active (-)								Condition
	Collector pump	Hot storage temperature higher than min. driving temperature	Thermal back-up	Sorption chiller	Hot water side valve	Chilled water side valve	Compression chiller	Load (approx.)	
Warm up solar circuit	+	-	-	-	-	(-)	-	0%	room air temperature and humidity o.k.; cold water is not needed
Compensate missing capacity in the morning	+	-	-	-	-	(-)	+	0 - 10%	cold water is needed but temperature of solar hot water is still too low
Solar cooling	+	+	-	+	+	(100%)	-	0 - 100%	solar hot water temperature is high enough and cold water temperature is low enough to cover the cooling load
Compensate missing capacity during peak load	+	+	-	+	100%	(+)	+	60 - 100%	solar hot water temperature is not high enough or cold water temperature is not low enough to cover the cooling load

Table 5.8
Operation scheme of a SHDC with an electrically driven chiller as back-up (example of Figure 5.7a)

5.2.3 Combined systems

A combined system includes a solar desiccant cooling system and a solar-thermally driven chiller (SDEC+SHDC systems). In such systems the advantages of both systems - SDEC and SHDC - are used to reach high energy efficiency. All these systems include a thermally driven chiller for chilled-water production and an air-handling unit with a desiccant wheel. Only collectors that use a liquid heat-transfer medium can be employed.

In this type of plant, the solar energy input is supplied to both the regeneration air for the desiccant wheel and the generator of the sorption chiller.

system number	liquid collector	air collector	heat storage	back-up heat	desiccant wheel	solar-thermally driven chiller	compression chiller	cold storage	description	application /example
5.3.1	X		X	X	X	X			thermally driven chiller and desiccant cooling system with solar collector and back-up heat source	strict requirements for indoor conditions in moderately humid climates/ e.g., conference room with high internal loads

Table 5.9
Classification of solar-assisted air-conditioning with combined SDEC and SHDC systems.

The advantage of this configuration is that the ventilation rate of the air-handling unit with solar desiccant cooling can be limited to the necessary fresh air requirements and does not have to be increased to match building loads. At the same time, the air dehumidification is effectively handled by the desiccant component without the need to drop below the dew-point, at least under most conditions, namely temperate ambient air humidity values. Therefore, the sorption chiller can be operated more efficiently, since a higher chilled-water temperature is sufficient and a higher cooling capacity can be generated. Consequently, the entire system has a higher energy efficiency and higher cooling loads can be removed. Systems employing conventional energy sources to completely supply either the desiccant regeneration component or the chiller generator are not covered in this section.

An overview of solar-assisted air-conditioning with combined systems is given in Table 5.9. Only one configuration is shown, which is the most feasible one from a technical point of view, given the current technological state of the art.

DESIGN OPTIONS

Depending on the weather conditions, combined system technology can be coupled to different cold distribution sub-systems. In particular, the SHDC can be used for providing chilled fluid to both a chilled-water distribution network, e.g., fan-coil systems, chilled ceilings, and a cooling coil installed in the air-handling unit. The selection among the possible configurations is based on the prevailing weather conditions and internal load profiles.

For mild conditions, the SDEC plant could be designed in order to completely cover the dehumidification needs and part of the sensible loads. The SHDC could then serve only the chilled-water distribution network for the building. Under more severe conditions, if the SDEC plant is not able to sufficiently cover the sensible loads, the circuit of the SHDC could serve a cooling coil integrated into the air-handling unit. In very humid climates, dehumidification can not be handled by the SDEC system alone, so the SHDC system may have to function as a dehumidification back-up. This situation has to be considered only in rare cases and should not be a normal operation mode. In fact, for very humid climates, a combined system will lose most of its advantages. Furthermore, a heat back-up is always necessary to ensure the operation of both systems; this means that solar-thermally autonomous systems are not considered further here, since the high effort regarding technical equipment seems not to be justified unless achieving comfort conditions can be always guaranteed. The back-up system should be integrated in such a way that control of both the regeneration temperature of the SDEC and the generator temperature in the SHDC is possible.

Theoretically, there are many possible configurations for combined systems. However, based on available practical experience, only one example is presented.

Example 5.3.1 **Solar-assisted system consisting of a thermally driven chiller and a desiccant cooling system; both sub-systems receive their heat from a solar collector and a back-up gas heater**

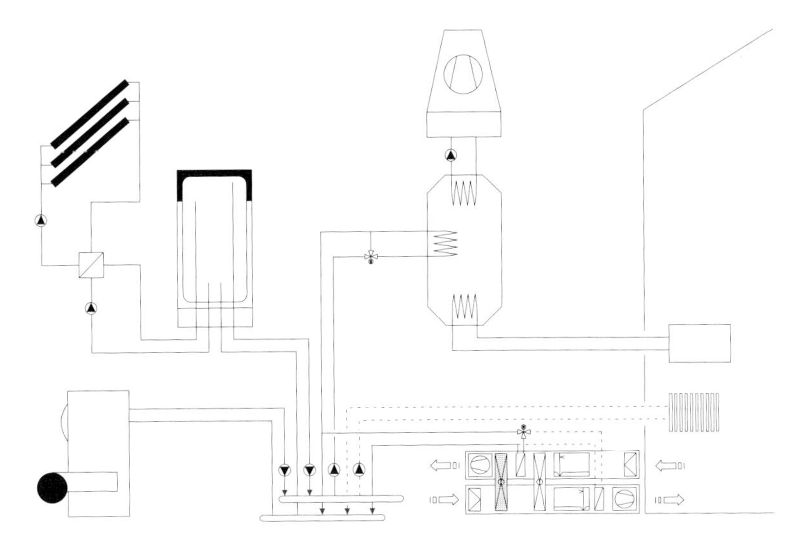

Figure 5.8
Solar-assisted system consisting of a thermally driven chiller and a desiccant cooling system; both sub-systems receive their heat from a solar collector and a back-up gas heater.

DESCRIPTION

Since the main components have already been extensively discussed in the previous sections, only a short description is given here for the most important components.

In contrast to the systems described in Section 5.1 and 5.2, in combined systems, the solar system has to provide energy to two energy-consuming systems, which may be connected in series or in parallel (as shown in Figure 5.8). Especially when sorption chiller driving temperatures higher than 80°C are used, the series connection is favourable. If the cooling load is small and can be covered by the SDEC system alone, the chiller generator is bypassed and only the regeneration heat exchanger is solar-thermally heated. The required SDEC driving temperature is a function of the load and in most cases is lower than the driving temperature of the sorption chiller. Therefore even in cases where the solar irradiation is not particularly high, it is likely to reach a sufficiently high temperature to drive the regeneration process of the SDEC.

During operation periods with abundant solar radiation, the sorption chiller is supplied with a high hot-water temperature, for example greater than 80°C, and the return flow, which is normally higher than 70°C, is used for the regeneration heat exchanger in the SDEC system. Depending on the actual conditions, the required water mass flow through the chiller will not be the same as the mass flow through the regeneration heat exchanger of the desiccant system. Therefore the lower of the two mass flows has to be adjusted using the respective bypass valve.

In cases where a lower driving temperature is used for the sorption chiller, e.g., for adsorption chillers, a parallel connection will be necessary, because the return temperature of the chiller generator is too low to supply the SDEC system. For both configurations - parallel (as in Figure 5.8) and series - a thermal back-up has to be used to heat the storage tank during periods when the solar irradiation is not high enough to meet the heat load of both systems.

In general, it should be noticed that a combined-system configuration requires more sophisticated technical equipment and thus will be more costly. Therefore, systems of this type are rarely installed at present. This configuration can only be recommended for large installations with a high overall cooling capacity.

CONTROL/ OPERATION

In general, the control and operation schemes presented in Sections 5.1 and 5.2 may be used as a basis for combined systems as well. Due to the large variety of possible configurations it is difficult to develop simple general control schemes.

6 DESIGN APPROACHES

In this chapter, different levels of the design process are described. In the context of this book, design mainly refers to the solar part of the whole technical system rather than to the design of air-handling units, chillers, heat exchangers or other components.

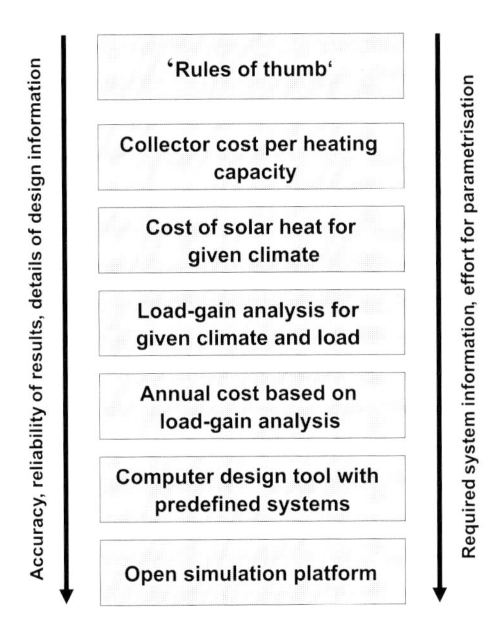

Figure 6.1
Overview of different design methods and levels.

Design of the solar system essentially means design of the solar collector system and the heat storage unit. However, the dimensioning of a cold water tank will also be influenced by the solar character of the heat source. In addition, the size of a back-up compression chiller depends also on the available cooling power from a solar thermally driven system. Similarly the size of a back-up heat source depends on the amount of solar heat available.

Different design methods are presented in this chapter starting with the simplest approach requiring a minimum of information and ending with references to more complex design tools and simulation programs.

DESIGN LEVELS

Different levels for the design of a solar collector system are possible, which are closely linked to the available meteorological and load data and which require quite different levels of effort.

Some of these methods will be described in detail in the following sections:

- Rules of thumb based on experience with installed systems, to be used as a starting point;
- Comparison of solar collectors with regard to the necessary investment per heating power at a design point;
- Comparison of solar collectors with regard to the cost of produced heat for a given climate and a given operation temperature;
- Design of a solar thermal system (collector, heat storage, back-up) for solar-assisted air-conditioning with regard to possible solar contributions to the overall energy demand for a given climate and a given load;
- Computer design tools which allow the simulation of a selected system on whole-year basis;
- More complex simulation tools, for example TRNSYS; only some general comments are made.

An overview of the different design levels is shown in Figure 6.1.

6.1 Rules of thumb

A very simple assessment of the collector dimensions in a solar-assisted air-conditioning system can be made using a single design point. Then the required collector area is defined by

$$A_{coll} = \frac{\dot{Q}_{low,\ design}}{G_{\perp} \cdot \eta_{design} \cdot COP_{thermal,\ design}} \quad [m^2] \qquad (6.1)$$

where $\dot{Q}_{low,\ design}$ is the nominal cooling capacity (kW), $G_{\perp}$ is the total solar radiation incident on the collector (kW), $COP_{thermal,design}$ is the coefficient of performance of the cooling system at design conditions, and η_{design} is the collector efficiency at design conditions.

The specific collector area, A_{spec}, which defines the collector area per nominal cooling capacity, is then defined as

$$A_{spec} = \frac{1}{G_{\perp} \cdot \eta_{design} \cdot COP_{thermal,\ design}} \quad \left[\frac{m^2}{kW_{cold}}\right] \qquad (6.2)$$

DESIGN POINT

For example, with an incident total solar radiation on the collector, $G_{\perp}$, of 800 W/m², a collector efficiency at design conditions of 50 % and a COP of the cooling system at design conditions of 65 %, the resulting specific collector area is about 3.8 m² per kW of cooling power.

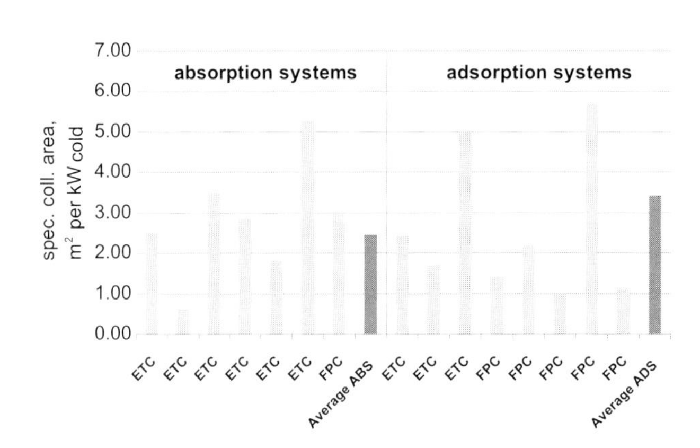

Figure 6.2
Collector size (absorber area) normalized to the installed chiller capacity for some absorption (ABS) and adsorption systems (ADS): ETC = evacuated tube collector; FPC = flat-plate collector.

This figure clearly provides only a first estimate of the collector area, since both the radiation and the cooling load can vary with time. Also the performance values, i.e., collector efficiency and COP, will change under part load conditions. In addition it does not make sense to dimension the solar collector for the maximum cooling load, since it might be overdimensioned for most of the time. If any back-up system is available, the annual solar fraction for the cooling load should typically be in the range of 70 - 80 % (see Chapter 7).

PRACTICAL EXPERIENCE

Based on available information from actual installations, it can be concluded that large differences in the installed collector area are possible. Representative data from several installations are shown in Figure 6.2 for systems that operate with absorption or adsorption chillers. The specific collector area varies between about 1 and 6 m² per kW of cooling power. This large range of values depends on the size of the solar installation and the overall strategy. For some cases, the solar array is used to supplement other heat sources, resulting in smaller values of the specific collector area. In addition, in some of the cases, the solar array is dimensioned to cover the total load for the cooling system and additional loads, for example, hot water heating. Based on the data shown in Figure 6.2, the average value for absorption systems is 2.5 m^2/kW_{cold} and is 3.4 m^2/kW_{cold} for adsorption systems.

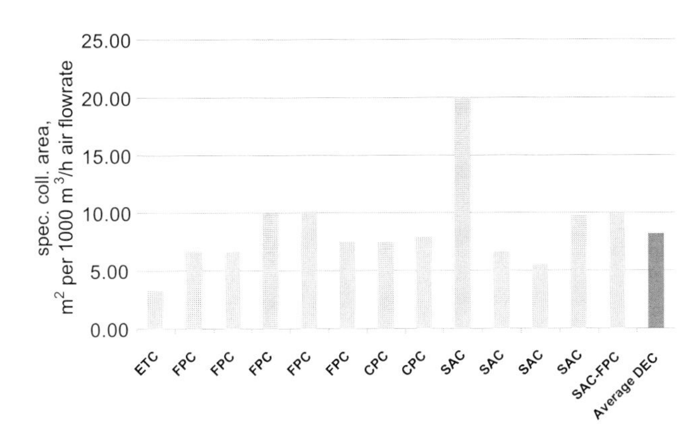

Figure 6.3
Collector size (absorber area) normalised to the installed nominal air flowrate for some desiccant-cooling systems; ETC = evacuated-tube collector; FPC = flat-plate collector; CPC = stationary CPC collector; SAC = solar air collector.

In Figure 6.3 similar values are shown for desiccant cooling systems. The specific collector area refers to the nominal air flowrate, since the definition of the cooling power strongly depends on the ambient and supply air conditions. The specific area for desiccant-cooling systems varies within a range of less than 5 m² per 1000 m³/h nominal air-flow and 20 m² per 1000 m³/h. The average value is about 8.2 m² per 1000 m³/h. In both cases, sorption chillers and desiccant-cooling systems, no clear correlation between the employed collector type, for example, evacuated-tube, flat-plate etc., and the installed specific collector area can be identified.

APPLICATION AND LIMITS

The figures given above can provide a first, very rough assessment of a typical collector area for a given system. The only information that is necessary is the size of the air-conditioning system or the thermally driven chiller, as applicable.

6.2 Comparison of solar collectors

MATHEMATICAL DESCRIPTION

At the beginning of a project, a designer needs to make an initial assessment of the feasibility of using solar energy and a rough estimate of the required collector area. A simple method that may be used to compare different solar collectors uses the efficiency parameters of the solar collectors and their specific cost values. The efficiency of a collector is calculated using Equation 4.3 in Section 4.1:

$$\eta = k(\Theta) \cdot c_0 - c_1 \cdot \frac{(T_{av} - T_{amb})}{G_\perp} - c_2 \cdot \frac{(T_{av} - T_{amb})^2}{G_\perp} \qquad (6.3)$$

where k(Θ) is the incident angle modifier, T_{av} is the average temperature in the collector (°C), T_{amb} is the ambient air temperature (°C), and $G_\perp$ the solar radiation incident on the collector (W/m²). c_0, c_1 and c_2 are collector efficiency values (see Section 4.1).

The method implies basing the efficiency calculation on typical values of the operation condition parameters, T_{av}, T_{amb} and $G_\perp$, in order to make the comparison between different solar collectors feasible. For the environmental parameters, T_{amb} and $G_\perp$, for instance average values of the 'main' hours of a typical clear sunny day during the cooling season can be used. For the collector operation temperature, i.e., the fluid average temperature, T_{av}, the driving temperature of the cooling equipment at nominal conditions can be used. The incident angle modifier, k(Θ), is set to 1 for this comparison.

The collector efficiency values are normally provided by manufacturers or can be retrieved from several commercial databases (e.g., /4.4/).

The flux of useful heat, $\dot{Q}_{use}$, of the collector is the product of the collector area, A, the collector's efficiency, η, and the normal incident solar radiation on the collector, $G_{\perp}$. Accordingly, the specific collector area, A_{spec}, which is needed to provide 1 kW of heat at assumed conditions, can be calculated as follows:

$$\dot{Q}_{use} = A \cdot \eta \cdot G_{\perp} \quad \Rightarrow \quad A = \frac{\dot{Q}_{use}}{\eta \cdot G_{\perp}} \quad \Rightarrow \quad A_{spec} = \frac{1kW}{\eta \cdot G_{\perp}} \qquad (6.4)$$

Multiplying the specific collector area with the specific collector price, $Cost_{spec}$, the investment cost per kW of heating power under design conditions can be estimated as follows:

$$Cost_{heat,\,power} = A_{spec} \cdot Cost_{spec} \qquad (6.5)$$

In the equation above the calculated $Cost_{heat,power}$ is expressed in €/kW and the $Cost_{spec}$ is expressed in €/m²; it is important to use the same reference collector area for efficiency and cost parameters in order to obtain coherent results.

Application Examples

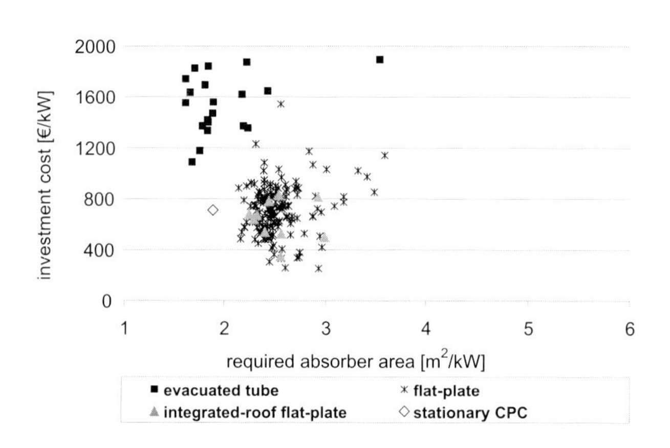

Figure 6.4
Specific investment cost for solar collectors versus required specific collector area. The values are valid for an ambient air temperature of 20°C, an average fluid temperature in the collector of 75°C and a radiation on the collector of 800 W/m².

The results of an analysis using the above method for different operating scenarios are shown in Figures 6.4 and 6.5. Figure 6.4 is valid for an ambient air temperature of 20°C, an average fluid temperature in the collector of 75°C and normal incident solar radiation on the collector of 800 W/m². These are typical operating conditions, for example, for a desiccant-cooling system or an adsorption chiller.

The lowest value of the specific cost (about 250 €/kW) is achieved with a flat-plate collector which requires a relatively large area for a heat production of 1 kW (about 3 to 3.3 m²). In general, roof-integrated flat-plate collectors also show comparatively low values of the cost figure, $Cost_{heat,power}$. The reason for this is that they are mounted in large single modules with areas of up to 12 m². This leads to a comparatively low cost per unit area. Figure 6.4 also shows that some types of solar collectors need quite large areas to supply 1 kW of heat at given conditions. These collectors have a low efficiency which has to be compensated by installing a larger area. Even if space is available to install them, this would be the wrong choice since the output from the collectors at lower solar radiation values would be much lower.

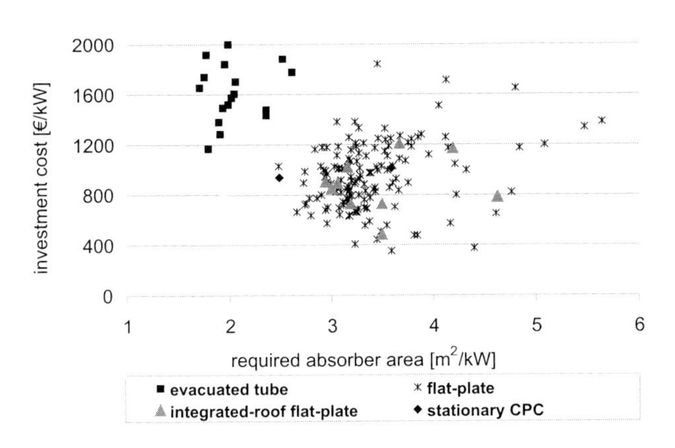

Figure 6.5
Specific investment cost for solar collectors versus required specific collector area. The values are valid for an ambient air temperature of 20°C, an average fluid temperature in the collector of 95°C and a radiation on the collector of 800 W/m².

A similar picture results for a less favourable condition, namely an average fluid temperature in the collector of 95°C and incident solar radiation on the collector of 800 W/m² (Figure 6.5). This refers for instance to operating conditions of a single-effect absorption chiller. The required collector area is larger compared to the scenario described before. This is particularly valid for the flat-plate collectors. Again, the same flat-plate collectors as before lead to the lowest specific cost value. However, the collector area required is more than two times larger than the one for the 'best' evacuated tube collector, i.e., the collector with the lowest required absorber area. Similar comparisons can be made with all parameters referring to the collector gross area. In this case the required area gives a better indication of the necessary roof space.

Application and Limits

The advantage of this method is that no information on the specific project is necessary, neither on the cooling load nor on the climatic data. The only required specific information is that of the operating temperature of the thermally driven cooling system at design conditions and the collector parameters. The result of this method is that a designer can exclude some types of collectors from further consideration at early stages of the design process. Also, a first estimate of the required collector area can be obtained, if the nominal heating power of the collector is known. Of course, this can only be a very rough estimate, since the incident solar radiation on the collector continuously changes during the day and throughout the year; so it is quite difficult to define the incident solar radiation at 'design conditions'.

6.3 Gross heat production of the collector

The next approach for selecting a solar collector for a given application is the calculation of the annual gross heat production, Q_{gross}, of the collector. This method is based on an annual time series of hourly values of solar radiation and hourly averages of the ambient air temperature. For each hour of the year, the useful heat that can be delivered by a particular collector with a defined orientation and tilt angle is calculated for a given operating temperature (average fluid temperature). The result of this method is a value for the maximum amount of heat that can be achieved by a selected collector at selected climatic conditions. The useful collector yield under real operating conditions is always below this value because generally not every unit of heat produced by the collector can be consumed by the application system. Only in the case of a very small solar fraction and a heat load which is much higher than the solar collector output throughout the year the actual collector output approaches the calculated gross heat production. In other words, the gross heat production of a solar collector can be considered as the collector yield in a system with a constant load temperature and an infinite load.

6.3.1 Meteorological data - climatic regions

In order to apply the collector gross heat production method, annual data sets of meteorological data are required. Data from seven reference locations have been selected which cover the whole range from the tropics to temperate-cold climates; these data sets were used to perform the calculations and generate the results presented in this handbook. The seven locations are:

- Merida, Mexico (20.42° north) represents a tropical climate with high humidity and temperature during the day and night, throughout the year.
- Athens, Greece (38.98° north) and Palermo, Italy (38.07° north) represent Mediterranean coastal climates with high humidity and temperature during summer and a moderate climate in winter. The typical cooling season is between April and October.
- Madrid, Spain (40.42° north) represents a Mediterranean, continental climate with high temperatures during summer but moderate air humidity. The heating season is longer than on the coast.
- Perpignan, France (42.7° north) represents a moderate Mediterranean climate.
- Freiburg, Germany (48.0° north) represents a warm, Central European climate.
- Copenhagen, Denmark (55.65° north) represents a moderate, Northern European climate.

METEOROLOGICAL DATA PROFILES

Figures 6.6 to 6.8 show the annual distributions of monthly averages of ambient air temperatures, relative humidity and global horizontal radiation totals, respectively, for each of the selected sites. The climatic data profiles have been extracted from a commercial database /6.1/. The average annual global horizontal radiation amounts to 988 kWh/m² in Copenhagen, 1115 kWh/m² in Freiburg, 1445 kWh/m² in Perpignan, 1566 kWh/m² in Athens, 1664 kWh/m² in Madrid, 1690 kWh/m² in Palermo and 1800 kWh/m² in Merida.

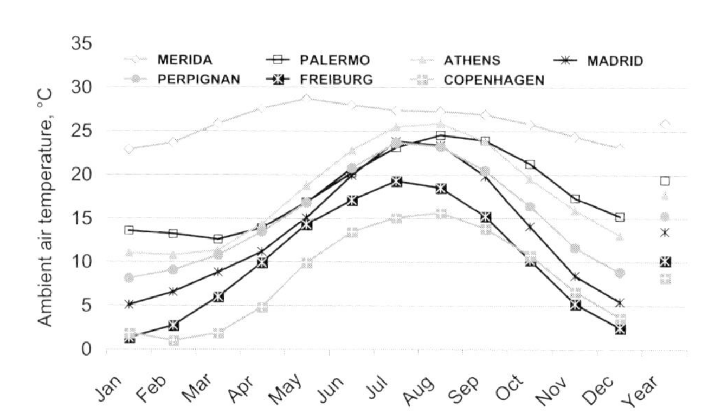

Figure 6.6
Annual distribution of average monthly ambient air temperatures for the seven locations used in this handbook.

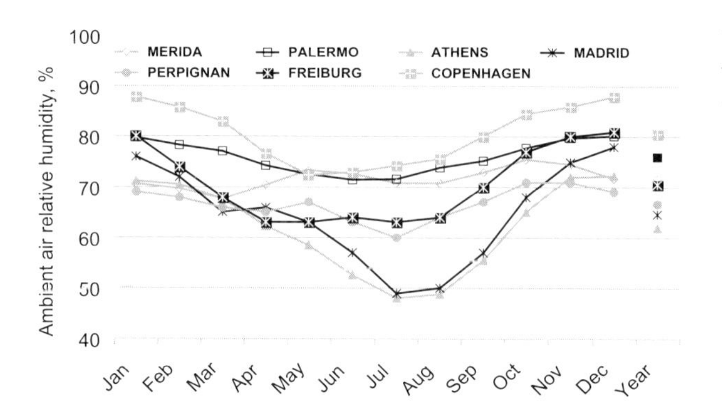

Figure 6.7
Annual distribution of average monthly ambient air relative humidity for the seven locations used in this handbook.

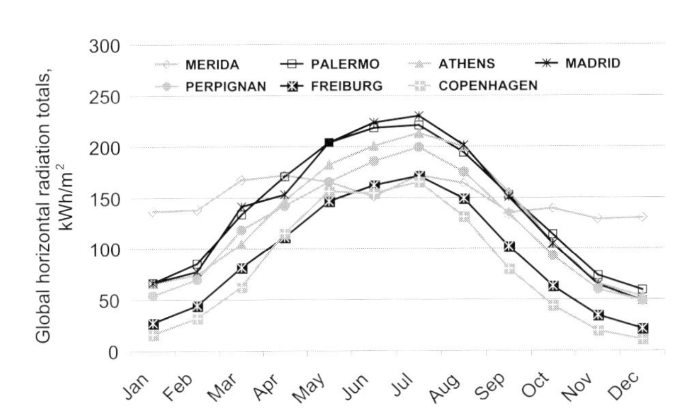

Figure 6.8
Annual distribution of average monthly global horizontal radiation totals for the seven locations used in this handbook.

6.3.2 Comparative results

The collector gross heat production, Q_{gross}, is calculated using the collector equation - see Equation 6.3 - taking into account the influence of the incidence angle (incident angle modifier) for each hour of the year. The collector operation temperature (the temperature of the fluid in the collector) is assumed constant. If the collector cannot deliver heat at this temperature, there is no contribution to the annual collector gross heat production for that hour. The calculation of the collector gross heat production is a standard feature of many software packages used for solar plant design (e.g., for solar domestic hot water systems /6.2/). For a selected solar collector type at a given location, the produced gross heat is calculated for each month of the year and summed up for the entire year to calculate the annual heat production.

An initial analysis of the specific cost of heat production can then be performed. The annual specific cost of the investment, $Cost_{annual}$, for the solar collector is the product of the specific collector cost, $Cost_{spec}$, and the annuity factor, $f_{annuity}$, and is calculated as follows:

$$Cost_{annual} = Cost_{spec} \cdot f_{annuity} \qquad (6.6)$$

The heat cost, $Cost_{heat}$, is then calculated by dividing the annual specific cost, $Cost_{annual}$, by the annual produced heat, Q_{gross}, as follows:

$$Cost_{heat} = \frac{Cost_{annual}}{Q_{gross}} \qquad (6.7)$$

COST OF SOLAR HEAT

The heat cost, $Cost_{heat}$, is expressed in €/kWh. Of course, this heat cost reflects only the investment for the solar collectors without taking into account any investment cost for the entire system and also without accounting for the operation cost, maintenance cost etc..

The results of an analysis using the above method for different scenarios, i.e., different types of solar collectors, operating conditions and locations, are shown in Figures 6.9 to 6.12. A value of 8.72 % has been used for the annuity factor, corresponding to an interest rate of 6 % and a lifetime of 20 years. The annual collector gross heat production and the specific heat cost are shown for two sites, namely for Freiburg (Germany) and Palermo (Italy), and for two operating temperatures and for a total of 188 commercially available solar collectors (23 evacuated tube collectors, 152 flat-plate collectors, 11 roof-integrated collectors and 2 stationary CPC collectors). The first operating temperature (75°C) is within the range of typically required driving temperature for desiccant-cooling systems. The second value (95°C) is typical for single-effect absorption chillers. In all calculations, it was assumed that the solar collectors are facing south, with a tilt angle of 30°. Based on the calculation results, the following conclusions can be drawn:

- The heat produced in southern Italy (Palermo) is nearly twice as high as the heat produced in central Europe (Freiburg).
- The average annual heat produced with evacuated tube solar collectors per absorber area is about twice as high as for flat-plate collectors. The values for stationary CPC collectors are in between.
- The cost of heat is lower with flat-plate collectors, roof-integrated systems and stationary CPC collectors, compared to the heat produced with evacuated tube collectors, even for high operating temperatures. These cost values include only the solar collector costs and do not take into account the cost for the solar collector's supporting structure and piping. Due to the larger collector area of flat-plate collectors needed to provide the same amount of heat, the cost difference decreases if all system costs are considered, particularly for high operating temperatures.
- The average heat cost for flat-plate collectors in Freiburg is about 8.5 €-cent/kWh at 75°C and 14.5 €-cent/kWh at 95°C. The corresponding figures for Palermo are 3.7 €–cent/kWh at 75°C and 5.5 €-cent/kWh at 95°C.
- The average heat cost for evacuated tube collectors in Freiburg is about 14 €–cent/kWh at 75°C and about 16.7 €–cent/kWh at 95°C. The corresponding figures for Palermo are 7.6 €-cent/kWh at 75°C and 8.6 €-cent/kWh at 95°C.
- The heat costs for roof-integrated collectors and stationary CPC collectors are the lowest and somewhat lower than average values for flat-plate collectors.

However, only two examples of stationary CPC collectors are considered in the comparison, so the results may not be representative.

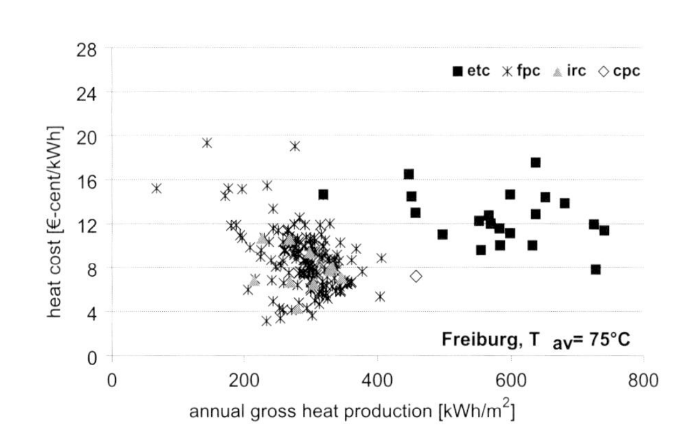

Figure 6.9
Heat cost versus annual collector gross heat production for different collector types (etc = evacuated tube collectors, fpc = flat-plate collectors, irc = flat-plate roof-integrated collectors, cpc = stationary CPC collectors) in Freiburg, Germany. The average fluid temperature is 75°C.

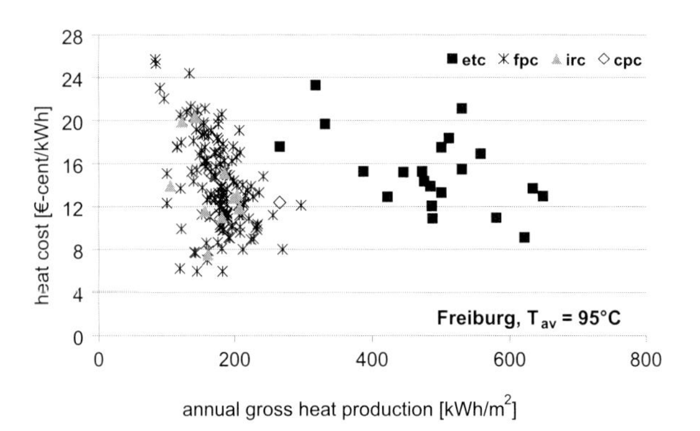

Figure 6.10
Heat cost versus annual collector gross heat production for different collector types (etc = evacuated tube collectors, fpc = flat-plate collectors, irc = flat-plate roof-integrated collectors, cpc = stationary CPC collectors) in Freiburg, Germany. The average fluid temperature is 95°C.

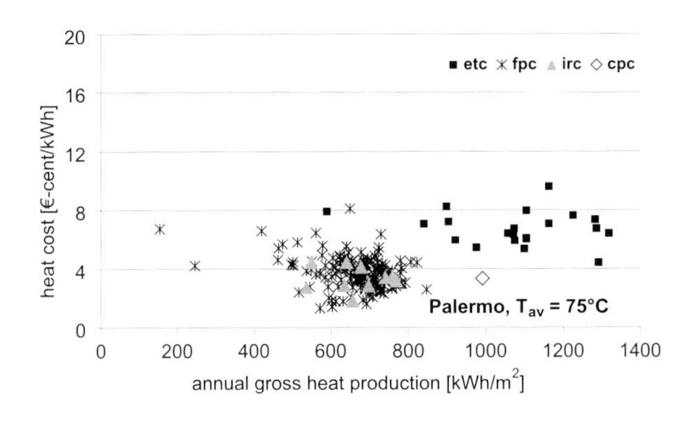

Figure 6.11
Heat cost versus annual collector gross heat production for different collector types (etc = evacuated tube collectors, fpc = flat-plate collectors, irc = flat-plate roof-integrated collectors, cpc = stationary CPC collectors) in Palermo, Italy. The average fluid temperature is 75°C.

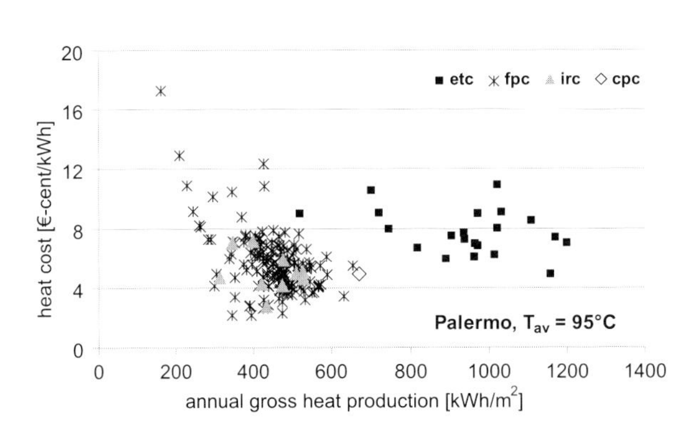

Figure 6.12
Heat cost versus annual collector gross heat production for different collector types (etc = evacuated tube collectors, fpc = flat-plate collectors, irc = flat-plate roof-integrated collectors, cpc = stationary CPC collectors) in Palermo, Italy. The average fluid temperature is 95°C.

6.4 Correlation between solar gains and cooling (heating) loads

METHOD DESCRIPTION

The method described in Section 6.3 is site specific because it is based on calculations using solar radiation data from specific locations. However, the influence of the load on the overall performance was not considered. To account for this, a further step can be implemented using the correlation between solar gains and cooling/heating loads on an hourly basis.

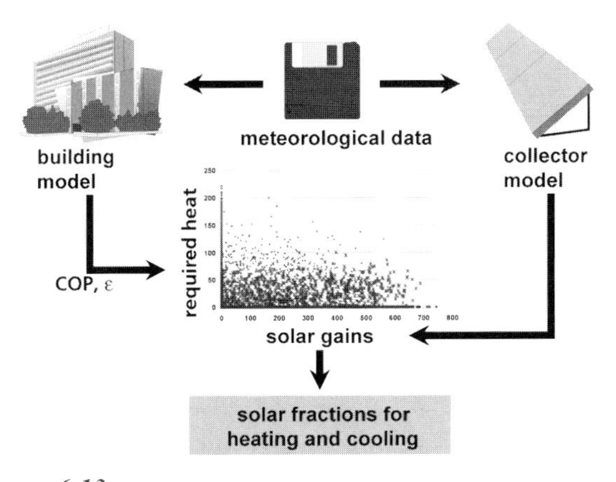

Figure 6.13
Schematic diagram of the method to analyse the correlation between solar gains and the heat required for cooling/heating.

This method is based on two time series: the time series of hourly values of solar gains for a given location (meteorological data) and a given operating temperature of the collector, and a time series of hourly values of the cooling or heating load of a given building. The hourly values of solar gains can be easily computed by using the collector equation as described in Section 6.2. The second time series needs a building simulation model for calculating the respective values, i.e., the heating/cooling loads for each hour of the year.

The building simulations can be done with various, currently available software packages. Among the most detailed are ESP-r (open public domain license), TRNSYS and EnergyPlus (commercial software); an extensive list of the software tools capable of producing the heating/cooling hourly series with different levels of detail, can be found in /6.3/. Based on the hourly values of cooling/heating loads, a simple assessment of the heat required to operate a thermally driven cooling system can be made by assuming a constant overall performance figure for the conversion of heat to cold. A general scheme of the method is shown in Figure 6.13.

6.4.1 Examples of loads

Different applications, for example, depending on the function of a building, can result in different loads. Three typical reference buildings have been selected to illustrate the corresponding loads. A detailed description is available in Appendix I. The three reference buildings are:

HOTEL

The reference hotel is a free-standing six-storey building. It is oriented along the north-south axis with an internal access corridor (in the centre of the building). The floor area of each storey, including the access areas, is 642.6 m^2. The glazed area on the north and south facade amounts to 25 % of the facades area and 4 % on the east and west facade. Load values were calculated for only one floor of the building.

OFFICE

The reference office is a free-standing, three-storey building with a basement. The building is oriented along the east-west axis with an internal access corridor in the center of the building. The floor area of each storey, including the access areas, is 309.9 m^2. The glazed area on the east facade amounts to 10 % of the facade area and about 37 % on the south and north facade. The west facade has no windows. Based on the dimensions of the building the characteristic length (= volume of building/area of building envelope) is 2.63 m.

LECTURE ROM

The lecture room represents a typical room where lectures, presentations, training courses or meetings are held. Here a human occupancy of around 2 m^2 per person corresponding to an internal load of 55 W/m^2 (latent and sensible heat) is assumed. The room has a simple rectangular geometry - 12 m wide by 18 m long - leading to a total floor area of 216 m^2. In comparison to the hotel and office building the lecture room is additionally characterised by a relatively high value for the minimal air exchange rate of 6.2 1/h (see Table 6.1), based on the minimal required air volume flow of 36 m^3 per person and hour.

The cooling load for the lecture room is dominated by ventilation loads due to the higher level of occupancy in comparison to the other two buildings. The cooling loads for the hotel and office buildings are dominated by internal loads and solar loads (due to solar gains). The operating hours for the office building are during the day, resulting in a good match of the peak cooling loads to the available solar radiation. On the other hand, for the hotel, the peak occupancy occurs during the evening. For each building, the construction, i.e., wall U-values, type of glazing etc., was defined according to typical building construction practice at the different reference climatic zones. Detailed information on the architecture, construction, internal loads etc. for each reference building is presented in Appendix I.

ANNUAL ENERGY FOR HEATING AND COOLING

The calculated annual energy and peak loads for cooling, heating, humidification and dehumidification are shown in Figure 6.14 for the three reference buildings and the seven locations. The loads account for all the internal and external building loads including infiltration and conditioning of ventilation air. The assumed infiltration and ventilation rates and desirable indoor conditions are given in Table 6.1.

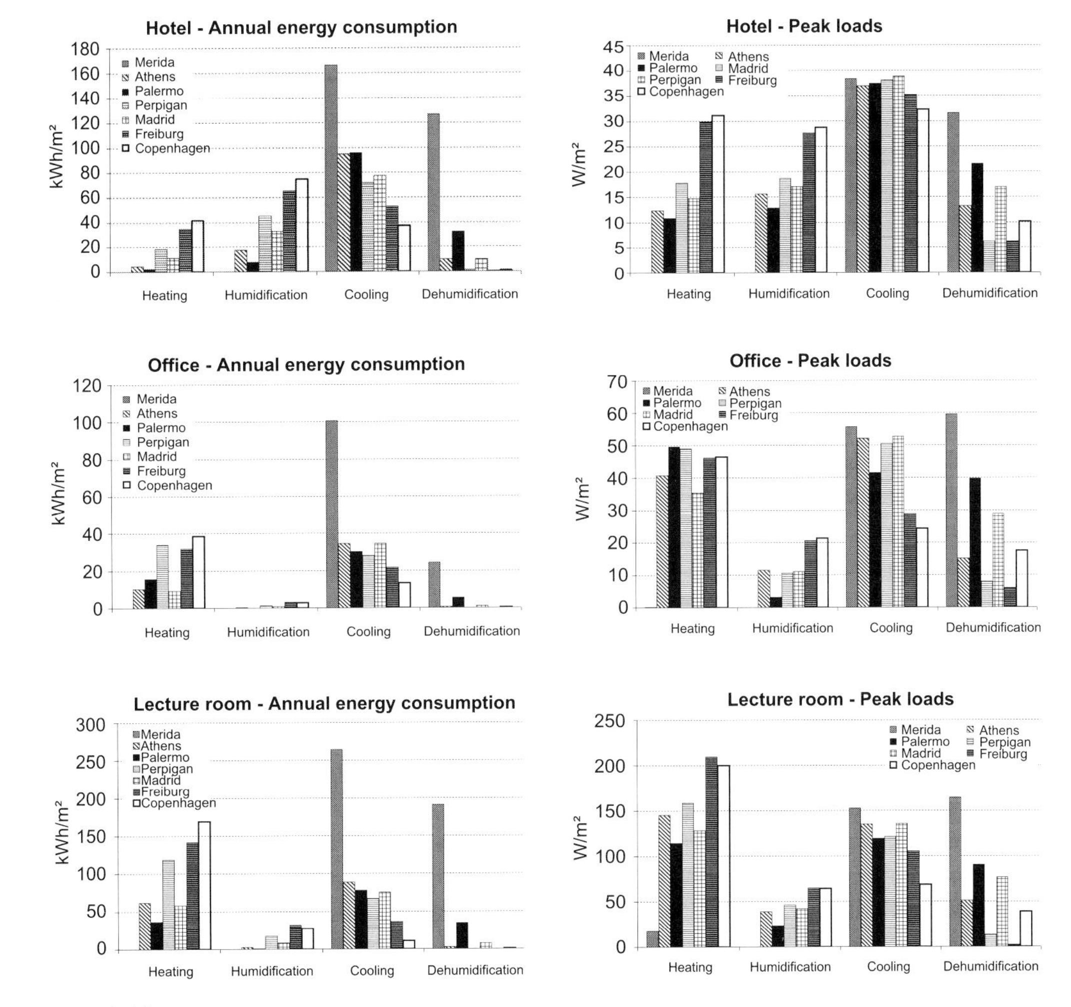

Figure 6.14
Annual heating, humidification, cooling and dehumidification energy consumption (left) and peak loads (right) for the different selected load examples (hotel, office and lecture room).

Parameter	Unit	Hotel	Office	Lecture room
Floor area	m^2	642.6	929.7	216
Air volume	m^3	1800	2956	648
Nominal ventilation rate during use (minimum required air-flow)	m^3/h	600	1000	4000
Air change rate due to ventilation	1/h	0.33	0.34	6.17
Infiltration rate	m^3/h	900	592	129.6
Air change rate due to infiltration	1/h	0.5	0.2	0.2
Total air change during use	1/h	0.83	0.54	6.37
Minimum indoor temperature	°C	22	20	20
Maximum indoor temperature	°C	26	26	26
Minimum indoor relative humidity	%	35	35	35
Maximum indoor relative humidity	%	50	60	60

Table 6.1
Air change conditions and comfort limits used to calculate heating/cooling loads of the reference buildings.

The highest energy consumption per unit floor area occurs in the lecture room due to the high ventilation rate associated with the high occupancy during use. The highest peak cooling and dehumidification loads occur in Merida (Mexico) because of the tropical weather conditions. Among the European sites, Palermo (Italy) shows the highest energy demand for dehumidification while Athens (Greece) has the highest demand for sensible cooling.

6.4.2 Selected results - solar fraction

APPLICATION EXAMPLE

The fraction of the total load which is covered by solar energy is referred to as the solar fraction, which is usually expressed as a percentage. As an example, the results of an analysis are shown for air-conditioning of an office building in Palermo. For this example the load file described in the previous section was used and the following parameters were selected: the supply temperature of the driving heat is 80°C, the effective COP of the cooling/air-conditioning system is 0.6, the temperature for heating is 45°C, the overall efficiency of the conventional heating system is 0.95, and the selected solar collector type is a stationary CPC collector. Further an adsorption chiller which produces chilled water was selected.

EXAMPLES OF RESULTS

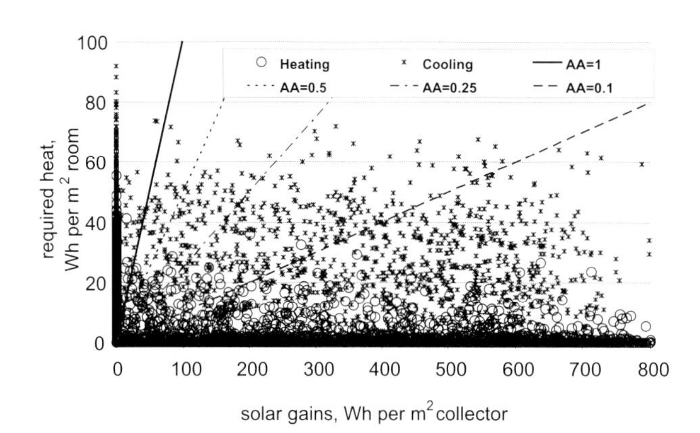

Figure 6.15
Comparison of hourly solar gains of a stationary CPC collector and heat required for cooling/heating an office building, using meteorological data for Palermo. The solid dots and crosses represent heating and cooling, respectively. The lines correspond to different AA-values that are defined as the specific collector area (m² of the solar collector absorber per m² of conditioned floor).

The calculation results are shown in Figure 6.15, illustrating the required heat for heating/cooling versus the available heat from the solar collector. Each point on the graph corresponds to a single hour of an entire year. The different lines shown in Figure 6.15 correspond to different specific collector areas, AA, defined as the installed solar collector absorber area per conditioned floor area. For a specific AA-line, a data point below the line indicates a situation in which the solar gains exceed the required heat. This excess heat is lost if the heat cannot be stored. Data points above the line indicate a situation in which solar gains are not sufficient to provide the heat required for cooling/heating. The difference between the available solar heat and the heat demand has to be delivered by another heat source. The annual solar fraction for cooling (SF_{cool}) can be calculated by summing up the hourly energy balance as follows:

$$SF_{cool} = 1 - \frac{Q_{bu\text{-}cool}}{Q_{tot\text{-}cool}} \tag{6.8}$$

where $Q_{bu\text{-}cool}$ is the annual back-up heat for driving the thermally driven cooling process, which is given by:

$$Q_{bu\text{-}cool} = \sum_{h=1}^{8760} Q_{bu\text{-}cool,\,h} \tag{6.9}$$

and where $Q_{tot\text{-}cool}$ is the annual required heat for driving the thermally driven cooling process, which is given by:

$$Q_{tot\text{-}cool} = \sum_{h=1}^{8760} Q_{tot\text{-}cool,\,h} \tag{6.10}$$

The hourly back-up heat for cooling, $Q_{bu\text{-}cool,h}$, is the difference between the heat required for cooling, $Q_{tot\text{-}cool,h}$, and the heat available from the solar collector, $Q_{coll,h}$, and is defined as follows:

$$Q_{bu\text{-}cool,\,h} = (Q_{tot\text{-}cool,\,h} - Q_{coll,\,h}) \tag{6.11}$$

where during the hours where the solar heat exceeds the hourly required heat for cooling, the correspondent $Q_{bu\text{-}cool,h}$ is set to zero.

The heat required for cooling, $Q_{tot\text{-}cool,h}$, is given by the cooling load, $Q_{cool\ load,h}$, divided by the COP of the cooling system, as follows:

$$Q_{tot\text{-}cool,\,h} = \frac{Q_{cool\ load,\,h}}{COP_{thermal}} \qquad (6.12)$$

Similar expressions can also be formulated for heating. The only difference is that the temperature for which the solar gains are calculated and the global efficiency during heating have different values.

STORAGE EFFECTS

The method can also be extended to study the effect of storage. For this purpose, a certain storage capacity is defined in terms of the maximum available energy content of the storage unit per collector area. If a storage tank is available, excess solar energy during a given hour can be used at a later time when solar gains are not sufficient to match the load. Solar excess heat is lost if the maximum storage capacity is achieved.

In a real system, storage can be integrated at different places within the system (see Section 4.2). Beside heat storage, a buffer could also be installed on the cold side, for example chilled water storage, or on the load side, for example employing thermal activation of building components.

DESIGN PARAMETER - SOLAR FRACTION

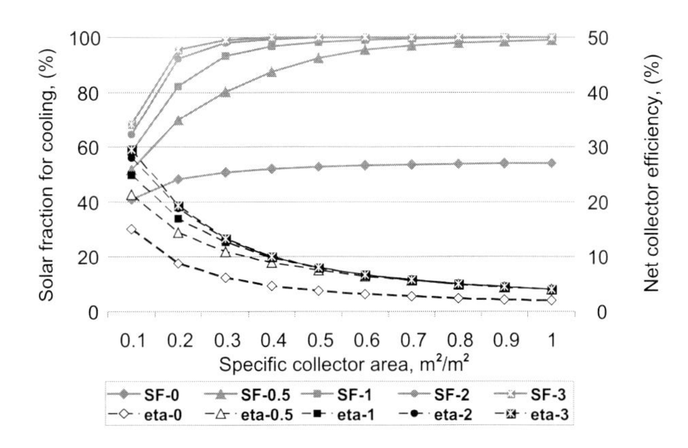

Figure 6.16
Solar fraction for cooling and net collector efficiency as functions of the specific collector area and heat storage dimensions (e.g. SF-1 = solar fraction for a heat storage unit with a capacity of 1 kWh/m² of collector; eta-1 = net collector efficiency for 1 kWh/m² heat storage capacity); the results are valid for the example of an office building in Palermo using a stationary CPC collector.

Using this method, a quick overview of possible solar fractions can be obtained for different collector areas, expressed in relation to the area of the conditioned floor and different values of the storage capacity. For the given example, the results are shown in Figure 6.16, presenting the solar fraction for cooling and the annual net collector efficiency. The net collector efficiency is defined as the fraction of useful solar heat with respect to the solar radiation incident on the collector. Figure 6.16 clearly shows that with increasing collector area, the solar fraction increases but the net collector efficiency decreases. Since the solar fraction approaches its maximum value (100 %), a further increase of the collector area above a certain value does not seem reasonable. For the given example, it turns out that a good compromise between a high solar fraction and acceptable net collector efficiency is achieved with a specific collector area of 0.1 to 0.2 m² of collector per m² floor area of conditioned space. The integration of heat storage is highly desirable. The solar fraction for cooling increases, for example, from about 48 % to more than 82 % for a specific collector area of AA = 0.2 m²/m², if a heat storage capacity of 1 kWh per m² of collector area is installed. If a temperature difference of 25°C (see Section 4.2.1) between storage maximum and minimum temperatures can be used, this storage capacity of 1 kWh corresponds to a hot water tank volume of about 0.035 m³ per m² of collector area. Using this storage volume, the net collector efficiency increases from about 8.5 % to more than 16 %. Further increase of the heat storage capacity leads to only small further increase of the solar fraction and the net collector efficiency.

APPLICATION PROFILE

The advantage of the correlation method is that it provides a good initial assessment of possible solar fractions and necessary collector area with a very limited effort. It does not involve any extensive simulation of the technical components, except the load (building) and the solar collector based on the stationary collector equation including consideration of the influence of the incidence angle. The method accounts correctly for the correlation of a specific load under given weather conditions with the available solar energy. However, the method is of limited accuracy, since it does not account for the part load conditions of any component in the chain of energy conversion, except for the solar collector and the load.

6.4.3 Parametric study

PARAMETRIC STUDY

Tables 6.3 to 6.5 present the results of a comparative study using the correlation method for the climates and loads described in Sections 6.3.1 and 6.4.1, respectively. The results include the annual solar fractions for cooling (Table 6.3) and heating (Table 6.4) and the annual net collector efficiency (Table 6.5). The parameters used for this study are summarised in Table 6.2.

Load		Lecture room	Hotel	Office building
Operation temperature, cooling	°C	70	90	80
Operation temperature, heating	°C	45	45	45
Global COP, cooling	-	0.7	0.65	0.6
Global efficiency, heating	-	0.95	0.95	0.95
Collector type	-	selective flat-plate	evacuated tube	stationary CPC
c_0	-	0.799	0.86	0.94
c_1	W/m^{2}K	3.12	2.02	2.2
c_2	W/m^{2}K^2	0.0119	0.0022	0.033

Table 6.2

Conditions for analysis of the solar fraction and net collector efficiency for different loads and meteorological data using the correlation method.

Solar Fraction Cooling		heat storage size in kWh per m^2 collector (absorber)																								
Load	Site	0					0.5					1					2					3				
		specific collector area, m^2 collector (absorber) per m^2 floor area																								
		0.1	0.2	0.3	0.5	0.8	0.1	0.2	0.3	0.5	0.8	0.1	0.2	0.3	0.5	0.8	0.1	0.2	0.3	0.5	0.8	0.1	0.2	0.3	0.5	0.8
Lecture room	Merida	9.5	18.7	26.5	38.2	47.4	10.0	19.7	28.7	43.5	58.4	10.4	20.6	30.2	46.4	64.1	11.2	22.1	32.6	50.1	69.4	11.9	23.6	34.6	53.3	73.6
	Athens	26.1	45.2	54.7	61.8	65.3	27.9	50.9	65.7	81.3	92.5	29.0	53.7	71.8	89.7	98.0	31.0	57.3	76.8	94.4	100.0	32.9	60.6	80.9	97.2	100.0
	Palermo	21.9	37.4	47.9	58.1	63.3	24.0	41.8	55.4	72.6	84.6	25.1	44.1	58.6	78.7	92.2	26.7	47.0	62.4	83.8	95.9	28.2	49.4	65.4	87.5	97.4
	Madrid	26.9	45.0	52.4	57.7	59.9	30.4	54.1	68.9	83.9	95.4	32.8	58.7	78.0	96.9	99.9	35.8	62.9	84.2	99.6	100.0	37.8	66.3	88.2	100.0	100.0
	Perpignan	22.0	36.5	45.0	52.6	56.4	25.2	44.6	57.8	75.2	86.8	27.2	48.5	64.7	83.6	93.1	29.6	52.6	70.1	87.8	95.8	31.4	55.9	73.8	90.3	97.2
	Freiburg	28.5	44.6	51.3	56.5	59.0	35.2	59.0	73.1	88.9	97.5	38.8	66.2	83.0	96.1	99.7	42.5	72.2	90.0	99.5	100.0	45.5	76.7	94.0	100.0	100.0
	Copenhagen	36.4	48.2	52.7	57.0	59.2	49.3	71.2	80.5	89.7	95.9	56.8	79.4	87.1	93.7	99.6	63.6	85.3	92.0	100.0	100.0	67.4	88.5	97.0	100.0	100.0
Hotel	Merida	20.5	28.0	30.8	33.1	34.5	21.7	34.2	41.2	51.5	64.3	22.2	38.8	49.9	66.8	85.0	22.4	43.5	60.2	81.5	94.7	22.4	44.2	62.5	84.6	97.5
	Athens	29.4	33.5	35.0	36.4	37.3	37.0	50.1	58.5	71.7	87.0	42.7	61.1	73.9	91.3	97.8	46.7	77.0	91.6	97.7	99.2	47.1	82.7	95.6	98.6	99.7
	Palermo	29.8	35.0	36.7	37.9	38.6	36.5	50.5	58.1	69.7	82.8	41.1	59.9	71.8	86.9	97.8	43.3	71.6	86.4	97.7	99.8	43.7	74.2	91.6	98.8	100.0
	Madrid	31.9	35.2	36.4	37.3	37.7	42.3	56.8	65.8	79.3	94.6	50.7	70.3	83.6	98.2	99.7	56.8	88.2	98.8	99.8	100.0	57.0	94.2	99.6	100.0	100.0
	Perpignan	27.8	32.4	33.9	35.5	36.3	35.7	50.5	59.4	72.7	87.1	41.1	60.8	73.8	89.9	96.6	44.0	73.0	88.8	96.6	99.2	44.2	76.2	93.0	98.4	100.0
	Freiburg	29.4	34.1	35.5	36.9	37.7	37.0	52.9	63.7	80.3	94.7	42.3	64.4	79.6	94.3	99.2	45.4	75.8	90.9	99.3	100.0	45.5	79.7	95.2	99.9	100.0
	Copenhagen	28.7	32.5	33.9	35.1	35.8	37.1	52.3	63.5	79.2	90.6	43.8	65.2	78.2	90.4	97.3	48.3	75.3	86.9	97.3	99.7	48.3	79.5	91.4	99.6	99.7
Office building	Merida	28.6	37.4	40.9	44.1	46.1	32.8	49.8	60.3	75.4	88.5	35.9	58.9	74.1	88.0	94.8	39.8	68.1	82.8	94.3	99.4	42.3	72.0	87.2	97.5	99.9
	Athens	41.7	47.7	49.7	51.5	52.9	53.1	70.4	81.4	92.6	95.8	61.3	84.9	92.7	96.5	98.6	70.3	93.0	96.9	99.1	100.0	74.8	95.9	98.3	100.0	100.0
	Palermo	41.0	48.2	50.7	52.7	53.8	51.7	69.7	80.1	92.5	98.0	58.4	82.2	93.3	98.2	99.8	64.7	92.2	97.9	100.0	100.0	68.3	95.5	99.0	100.0	100.0
	Madrid	46.9	54.4	57.0	59.6	61.2	58.8	76.7	87.1	96.5	98.3	66.0	90.1	97.2	98.6	99.5	73.5	97.3	98.9	99.9	100.0	77.4	98.7	99.7	100.0	100.0
	Perpignan	40.8	47.6	50.2	52.7	54.0	52.0	69.6	79.7	89.8	94.7	58.4	81.0	89.9	95.5	98.1	64.7	90.4	95.7	98.9	100.0	68.2	94.0	97.9	99.8	100.0
	Freiburg	42.8	50.7	53.7	55.7	57.0	54.7	74.4	84.2	94.0	99.2	60.0	81.8	92.3	99.2	100.0	66.8	90.9	99.1	100.0	100.0	71.3	97.1	100.0	100.0	100.0
	Copenhagen	46.8	52.1	54.4	56.3	57.3	60.9	77.4	84.7	93.5	98.5	67.2	85.2	93.0	98.8	100.0	73.7	93.9	99.0	100.0	100.0	78.9	98.3	100.0	100.0	100.0

Table 6.3

Solar fraction for cooling as a function of specific collector area and storage capacity (kWh per m^2 of collector) for the different loads and sites. The values are valid for the system parameters described in Table 6.2.

Since a lecture room needs a larger amount of ventilation air, the parameters that correspond to a desiccant-cooling system were used. The system parameters used for the hotel building correspond to an absorption chiller and for the office building, they correspond to the parameters of an adsorption chiller. The selected types of the solar collector are a flat-plate collector for the seminar room, an evacuated tube collector for the hotel and a stationary CPC collector for the office building.

Solar Fraction Heating		heat storage size in kWh per m^2 collector (absorber)																								
		0					0.5					1					2					3				
Load	Site	specific collector area, m^2 collector (absorber) per m^2 floor area																								
		0.1	0.2	0.3	0.5	0.8	0.1	0.2	0.3	0.5	0.8	0.1	0.2	0.3	0.5	0.8	0.1	0.2	0.3	0.5	0.8	0.1	0.2	0.3	0.5	0.8
Lecture room	Merida	-	-	-	-	-	-	-	-	-	-	-	-	-	-	-	-	-	-	-	-	-	-	-	-	-
	Athens	14.0	19.6	22.4	25.4	27.5	21.5	35.4	45.5	58.5	71.3	26.2	43.4	54.5	69.2	81.9	30.8	49.2	61.7	77.9	90.8	33.1	52.5	66.2	83.0	93.8
	Palermo	20.8	27.5	31.0	34.7	37.1	36.4	57.3	70.2	84.4	93.4	48.3	72.5	84.1	93.7	98.4	58.2	83.9	93.2	98.8	99.7	62.5	87.9	97.6	99.7	99.7
	Madrid	10.5	16.4	19.7	23.1	25.5	14.2	25.1	33.9	47.1	61.1	16.9	31.1	42.3	57.9	72.6	20.0	36.7	48.1	63.8	80.1	21.1	38.6	50.7	66.5	83.1
	Perpignan	12.1	17.5	20.2	23.0	25.0	18.0	32.1	44.0	59.9	75.3	22.4	42.2	55.9	73.1	86.5	28.0	48.9	63.5	81.6	92.3	30.0	51.7	67.4	85.2	96.9
	Freiburg	4.3	7.3	9.4	11.9	13.8	5.9	10.9	15.2	22.2	29.8	6.9	13.2	18.1	26.2	35.3	8.0	15.0	20.5	29.1	39.3	8.3	15.9	21.4	30.0	40.9
	Copenhagen	5.1	8.4	10.8	13.6	15.9	7.1	12.4	16.6	23.4	30.9	8.3	14.5	19.2	26.4	34.6	9.6	16.5	21.6	29.1	38.0	10.2	17.6	22.7	30.7	39.5
Hotel	Merida	-	-	-	-	-	-	-	-	-	-	-	-	-	-	-	-	-	-	-	-	-	-	-	-	-
	Athens	16.2	18.2	18.7	19.3	19.7	23.5	34.8	55.4	83.4	95.6	28.8	64.0	83.5	96.4	98.5	45.5	83.2	95.0	99.2	99.5	49.3	89.1	98.1	99.5	99.5
	Palermo	16.0	17.5	18.0	18.4	18.7	23.3	37.7	72.9	96.0	99.4	32.7	83.3	98.7	99.4	99.4	69.5	99.1	99.4	99.4	99.4	77.6	99.4	99.4	99.4	99.4
	Madrid	16.3	18.3	19.2	19.8	20.2	21.9	31.7	45.1	71.3	91.9	26.4	52.1	75.5	93.1	97.4	36.6	74.8	89.8	97.6	99.7	39.3	79.3	92.5	99.3	99.7
	Perpignan	15.9	17.6	18.3	18.8	19.1	22.1	31.8	47.0	76.7	93.9	27.0	55.5	78.7	94.2	98.0	38.0	77.3	92.1	98.8	99.7	39.6	81.0	95.7	99.7	99.7
	Freiburg	9.8	12.1	13.0	13.9	14.6	12.4	19.2	25.9	40.0	57.0	13.9	26.6	39.0	55.4	68.1	15.7	34.9	47.8	63.8	76.4	16.0	36.3	49.9	67.2	80.6
	Copenhagen	11.1	13.6	14.6	15.6	16.3	13.7	21.0	27.6	41.8	56.6	15.2	27.7	40.6	54.5	66.2	17.0	35.1	46.8	60.3	72.9	17.5	36.9	48.9	63.0	75.8
Office building	Merida	-	-	-	-	-	-	-	-	-	-	-	-	-	-	-	-	-	-	-	-	-	-	-	-	-
	Athens	21.8	22.9	23.5	24.0	24.4	60.5	77.4	85.6	93.1	96.9	72.1	87.4	94.1	97.8	99.7	82.2	96.1	98.7	99.7	99.7	86.6	98.7	99.7	99.7	99.7
	Palermo	24.7	26.9	27.8	28.6	29.1	59.1	79.3	89.0	97.3	99.8	75.2	91.7	97.9	99.8	99.8	87.7	99.0	99.8	99.8	99.8	92.5	99.8	99.8	99.8	99.8
	Madrid	16.2	17.6	18.1	18.6	19.1	36.8	55.1	67.5	83.0	91.8	49.7	71.6	83.9	92.7	97.1	60.2	82.3	90.3	97.9	99.8	63.4	85.8	93.7	99.8	99.8
	Perpignan	21.5	22.7	23.1	23.4	23.5	63.6	84.2	92.3	96.6	99.5	80.3	94.1	97.6	99.7	99.7	90.8	99.0	99.7	99.7	99.7	94.8	99.7	99.7	99.7	99.7
	Freiburg	8.6	9.4	9.6	9.8	10.0	20.1	31.6	39.8	51.3	62.3	26.4	41.2	50.0	62.3	73.9	31.1	48.0	59.2	73.1	81.8	32.1	51.1	65.4	77.7	84.6
	Copenhagen	9.8	10.7	11.0	11.3	11.4	20.9	31.2	38.6	50.2	60.8	25.7	38.3	47.2	59.4	69.2	28.7	43.1	53.5	66.9	77.0	29.6	44.9	56.5	71.5	79.7

Table 6.4

Solar fraction for heating as a function of the specific collector area and storage capacity (kWh per m^2 of collector) for the different loads and sites. The values are valid for the system parameters described in Table 6.2.

Net Collector Efficiency		heat storage size in kWh per m^2 collector (absorber)																								
		0					0.5					1					2					3				
	Site	specific collector area, m^2 collector (absorber) per m^2 floor area																								
		0.1	0.2	0.3	0.5	0.8	0.1	0.2	0.3	0.5	0.8	0.1	0.2	0.3	0.5	0.8	0.1	0.2	0.3	0.5	0.8	0.1	0.2	0.3	0.5	0.8
Lecture room	Merida	32.9	32.2	30.5	26.3	20.4	34.4	34.0	33.0	30.0	25.2	35.7	35.4	34.7	32.0	27.6	38.5	38.2	37.4	34.6	29.9	41.0	40.7	39.8	36.8	31.7
	Athens	25.5	21.1	16.9	11.4	7.6	29.8	26.4	22.7	17.0	12.3	32.5	29.1	25.4	19.2	13.4	35.9	31.6	27.7	20.6	14.0	38.2	33.5	29.3	21.4	14.2
	Palermo	23.0	18.8	15.7	11.3	7.7	28.0	23.7	20.6	15.9	11.4	31.3	26.3	22.4	17.3	12.4	34.8	28.7	24.1	18.4	12.8	36.9	30.1	25.3	19.0	13.0
	Madrid	23.9	19.5	15.3	10.3	6.9	28.9	25.7	22.3	17.3	13.1	32.5	29.5	26.4	20.5	14.6	36.7	33.0	29.1	21.8	15.4	38.8	34.8	30.6	22.3	15.7
	Perpignan	22.1	17.8	14.4	10.0	6.8	27.1	24.1	21.2	16.8	12.4	30.7	27.8	24.7	19.2	13.7	35.0	30.9	27.2	20.6	14.2	37.3	32.8	28.7	21.3	14.6
	Freiburg	18.7	15.1	12.1	8.5	5.8	23.9	20.9	18.1	14.3	10.9	27.1	24.2	21.0	16.2	12.0	30.3	26.8	23.2	17.3	12.8	32.1	28.5	24.2	17.7	13.1
	Copenhagen	14.4	11.0	8.9	6.5	4.6	19.8	16.2	13.8	11.0	8.6	23.0	18.6	15.7	12.2	9.6	26.2	20.9	17.3	13.3	10.3	27.9	22.1	18.2	13.9	10.7
Hotel	Merida	48.7	33.3	24.4	15.8	10.3	51.7	40.7	32.7	24.6	19.1	52.8	46.2	39.6	31.8	25.3	53.3	51.9	47.8	38.8	28.2	53.3	52.7	49.6	40.3	29.0
	Athens	29.9	17.0	11.8	7.4	4.7	38.1	26.0	20.9	15.8	11.9	44.1	33.1	27.0	19.8	13.2	50.2	41.9	33.1	21.1	13.4	51.1	45.1	34.5	21.3	13.4
	Palermo	32.2	18.9	13.2	8.2	5.2	39.6	27.5	21.7	15.7	11.5	45.0	33.7	26.9	19.3	13.5	49.2	40.3	32.1	21.6	13.8	50.2	41.7	33.9	21.8	13.8
	Madrid	27.4	15.1	10.5	6.5	4.1	36.4	24.8	20.3	16.0	12.3	43.6	33.3	28.2	20.2	13.0	51.6	43.7	33.4	20.8	13.1	52.9	46.5	33.9	20.9	13.1
	Perpignan	28.7	16.5	11.5	7.2	4.6	37.2	26.4	21.7	17.0	12.8	43.3	34.4	28.9	21.0	14.0	49.1	42.8	34.5	22.5	14.4	49.8	44.8	36.1	22.8	14.4
	Freiburg	28.7	17.0	11.9	7.5	4.8	36.2	26.5	22.1	18.0	14.3	41.1	33.7	29.5	22.6	15.9	44.8	41.2	34.7	24.8	16.9	45.1	43.2	36.3	25.5	17.4
	Copenhagen	26.6	15.7	11.0	7.0	4.5	33.8	24.7	20.7	17.1	13.4	38.8	31.6	27.8	21.0	15.1	43.0	38.1	31.6	22.9	16.1	43.6	40.2	33.1	23.7	16.5
Office building	Merida	31.5	20.6	15.0	9.7	6.3	36.2	27.4	22.1	16.6	12.2	39.6	32.5	27.2	19.4	13.1	43.9	37.5	30.4	20.8	13.7	46.6	39.7	32.0	21.5	13.8
	Athens	15.6	8.8	6.1	3.8	2.4	21.9	14.4	11.0	7.5	4.8	25.4	17.2	12.5	7.8	5.0	29.1	18.9	13.1	8.0	5.0	31.0	19.5	13.3	8.1	5.0
	Palermo	15.0	8.7	6.1	3.8	2.4	21.4	14.4	11.0	7.5	4.9	24.9	16.9	12.6	7.9	5.0	28.0	18.8	13.2	8.0	5.0	29.6	19.4	13.3	8.0	5.0
	Madrid	16.4	9.4	6.5	4.1	2.6	24.1	16.5	12.8	8.9	5.9	28.9	20.1	15.0	9.4	6.0	33.2	22.3	15.6	9.7	6.1	35.0	22.8	15.9	9.8	6.1
	Perpignan	16.8	9.7	6.8	4.3	2.7	23.9	16.0	12.1	8.1	5.3	27.4	18.4	13.5	8.5	5.5	30.5	20.4	14.3	8.8	5.5	32.1	21.1	14.5	8.9	5.6
	Freiburg	15.4	9.0	6.3	3.9	2.5	22.5	15.9	12.4	8.8	6.1	26.0	18.5	14.3	9.7	6.6	29.5	20.9	15.9	10.4	6.9	31.1	22.3	16.6	10.7	7.0
	Copenhagen	12.8	7.1	4.9	3.0	1.9	19.8	13.5	10.4	7.4	5.3	22.8	15.6	12.0	8.4	5.7	25.2	17.3	13.2	9.0	6.1	26.6	18.1	13.6	9.3	6.2

Table 6.5

Net collector efficiency as a function of the specific collector area and storage capacity (kWh per m^2 of collector) for the different loads and sites. The values are valid for the system parameters described in Table 6.2.

CONCLUSIONS

Based on the calculation results, the following conclusions can be drawn:

- In all cases, the integration of any type of storage is highly recommended since both the solar fraction and the net collector efficiency will increase; the recommended storage capacity is about 1 kWh per m^2 of collector area for most sites.
- For the office building, a specific collector area of 0.2 m^2 per m^2 of conditioned area leads to solar fractions for cooling which are higher than 80 % in all locations, except for Merida, assuming that storage is also used (1 kWh per m^2). Due to the high latent loads in Merida, larger collector areas are required in order to achieve the same solar fraction.
- For the hotel building, either a larger storage capacity (2 kWh per m^2) or a larger collector array (0.3 m^2 per m^2 conditioned floor area) is necessary, in order to achieve solar fractions higher than 70 %. Again the situation looks different in Merida, where a solar fraction greater than 70 % requires for instance a specific collector area of 0.5 m^2/m^2 and a storage capacity of 2 kWh/m^2.
- For the lecture room larger collector areas are required at almost all sites due to the much higher internal loads.
- In all southern European climates, a large fraction of the heating load can also be covered using solar energy. In most cases, the solar fraction for heating is about the same as the solar fraction for cooling. In this way, the use of the solar collector is extended throughout the year, compared to its use for air-conditioning in summer only. In Freiburg and Copenhagen, the solar fraction for heating is appreciably lower than in the Mediterranean locations. In Merida there is no demand for heating.
- Even if storage is used, the net collector output is very much lower than the maximum possible annual value, i.e., the annual gross heat production as calculated using the method described in Section 6.3. The difference is particularly large, if no or only a small storage unit is used. The reason is that for many hours during a day when solar radiation is available, there may be no demand for heating or cooling, for example, during spring and autumn.

6.5 SACE method

A more detailed design method, based on the one previously presented, has been developed within the SACE project (Solar Air-Conditioning in Europe, funded by the European Commission). In comparison to the approach described in Section 6.4. (storage in terms of available energy per collector area), a simple physical model of the heat/cold storage unit has been implemented. Also, a distinction between all air systems, water systems and air-water systems has been introduced in order to allow consideration of heat recovery during heating and variable volume flow. Finally, simple models for additional components such as pumps, cooling tower, compression chiller and heat exchanger have been introduced in order to obtain a more accurate picture of the overall energy balance.

A total of eleven different systems have been pre-defined, of which seven systems use solar energy and four are reference systems using conventional heating and cooling equipment. An overview of the different systems is given in Table 6.6 and Figure 6.17.

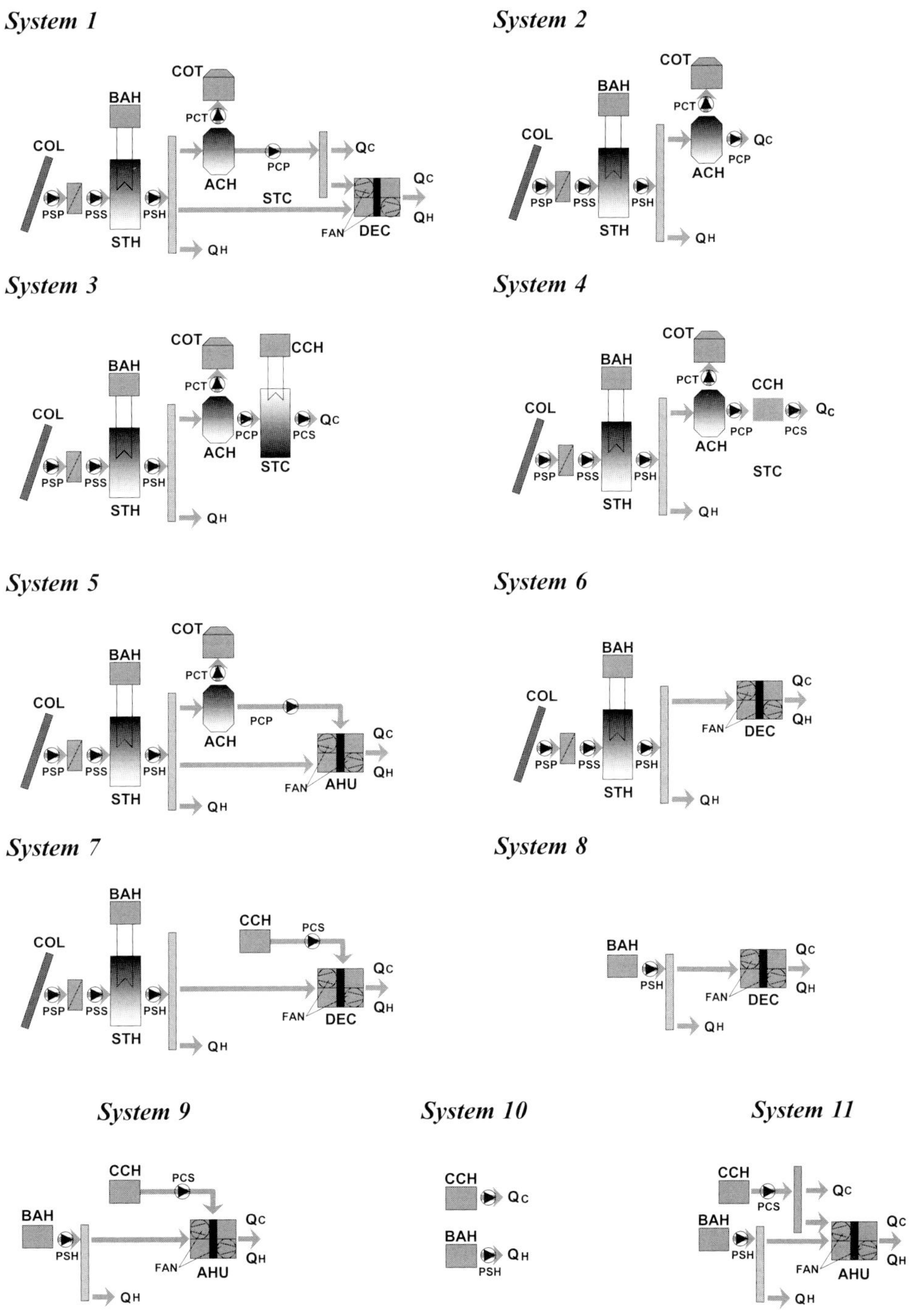

System 1 ***System 2***

System 3 ***System 4***

System 5 ***System 6***

System 7 ***System 8***

System 9 ***System 10*** ***System 11***

Figure 6.17

Configurations of systems in the SACE model; meaning of abbreviations:

COL	*solar collector*	*CCH*	*compression chiller*
ACH	*thermally driven chiller*	*COT*	*cooling tower*
BAH	*conventional (back-up) heater*	*PSP*	*primary pump solar*
AHU	*air-handling unit (conventional)*	*PSS*	*secondary pump solar*
DEC	*desiccant air-handling unit*	*PSH*	*heating cycle pump*
STH	*heat storage tank (hot water)*	*PCP*	*primary cold cycle pump*
STC	*cold storage tank (chilled water)*	*PCS*	*secondary cold cycle pump*
QH	*heat load*	*QC*	*cooling load*

METHOD

The method is based on the one described in the previous section. For a specific application (i.e., typology of load at certain site), the hourly heating/cooling load profiles must be available in a load file. Based on the load data the required heat for thermally driven cooling or for heating is calculated. Based on the collector parameters and the actual meteorological data, the output of the solar collector is calculated. The operation temperature for the collector is determined by the actual heat storage temperature. Depending on the chosen type of back-up system, either a back-up heat source or an electrical chiller is employed, if neither sufficient solar heat nor heat in the heat buffer is available. In case of higher solar gains than needs, the excess heat is stored in the buffer unit. However, the capacity is limited according to the maximum possible temperature of the heat buffer storage. Depending on which components are active, the corresponding pumps are identified and their electricity consumption is summed up. The output of this method is a complete annual energy balance including an economic analysis. The results of a parametric study performed using this method are described in Section 7.2.1. More results can for instance be found in /6.4, 6.5/.

system no	solar collector	heat storage	backup heater	thermal chiller	cold storage	compr. chiller	air-handling unit	desiccant	system description
1	+	+	+	+			+	+	air-water system; solar heat for chiller and desiccant system; back-up heater for heating and cooling
2	+	+	+	+					water-system; thermal chiller driven by solar heat; back-up heater for cooling and heating
3	+	+	(+)	+	+	+			water system; thermal chiller driven by solar heat; back-up cooling by compression chiller; backup heater for heating only
4	+	+	(+)	+		+			like system 3 but without cold water storage
5	+	+	+	+			+		air system; thermal chiller driven by solar heat coupled to air-handling unit; back-up heater for cooling and heating
6	+	+	+				+	+	air system; desiccant cooling system driven by solar heat; back-up heater for cooling and heating
7	+	+	(+)			+	+	+	air system; desiccant cooling system driven by solar heat; back-up cooling by compression chiller; backup heater for heating only
8			+				+	+	air system; desiccant cooling system completely driven by conventional heater
9			(+)			+	+		air system; conventional air-handling unit with compression chiller for cooling
10			(+)			+			water system; compression chiller for cooling; heater for heating
11			(+)			+	+		air-water system; compression chiller for chilled water directly and for the air-handling unit; heater for heating

(+) means: backup heater only for heating, not for cooling

Table 6.6

Overview on the systems available in the SACE model.

APPLICATION PROFILE

The advantage of the method is that it provides a more complete picture of the entire system than the correlation method alone, since all major components that contribute to the system energy performance are considered. The main limitation is that only global performance figures for each component are used except for the solar collector and the load. The software allows for a quick parametric study of different systems. For example, the annual simulation of 50 system configurations does not take more than 10 seconds on a personal computer with a 1.8 GHz Pentium processor. Thus, the method allows a quick first assessment of the optimum heat storage size, collector type and size for a given application at a given site. The method turns out to be suitable for feasibility studies at early stage of a building project.

6.6 TASK 25 design tool

OBJECTIVE

When deciding on a specific solar-assisted air-conditioning system, it is necessary to use simulation tools due to the complexity of the systems. Although several established simulation programs for solar thermal systems are available (see Section 6.7), a new software was developed within the framework of the IEA SHC Task 25. The focus was on user-friendliness, so that the user could work with the program without extensive prior training.

The simulation program calculates the hourly energy demand of the components in a solar-assisted air-conditioning system which consists of:

- Solar energy supply.
- Refrigeration.
- Air-handling unit.
- Room components.

The calculations are performed on an hourly basis and are summed up to calculate the annual energy demand. The building loads (e.g., calculated using the method for the energy demand of buildings specified in the German standard VDI 2067, Sheet 10 and 11 or any other building simulation program like TRNSYS, EnergyPlus, ESP-r etc.) do not account for any effects resulting from insufficient performance of the refrigeration and air-conditioning units. The annual total costs of the system are calculated based on the annual energy demand of the components and their investment, maintenance and capital costs. The purpose of this simulation program is primarily to quickly calculate the main system parameters in order to find the optimal solution for a specific application.

6.6.1 Method and software structure

CALCULATION APPROACH

Pursuant to the objective, a simulation program was developed which supports a user during the entire designing and decision-making process when dealing with solar air-conditioning. Figure 6.18 gives an overview of the software's structure. When designing the program, particular attention was paid to detailed calculation of the performance capability of the solar energy supply. In the course of the hourly simulation of the entire system, one calculation is implemented for determining the demand of refrigeration and air-conditioning (forward calculation) and a subsequent calculation is made - based on the results of the solar energy supply (backward calculation) - for determining the actual performance and operation conditions of the components. Thus, faulty dimensioning of the components becomes apparent, or the effect of deliberate under-sizing, with simultaneous acceptance of a certain, temporary non-compliance with the fresh air specifications, can be estimated. As reliable statements regarding the components to be used can hardly be made in the early phase of deciding on the system, the input information was limited to the necessary minimum and to the data that are commonly available at this stage.

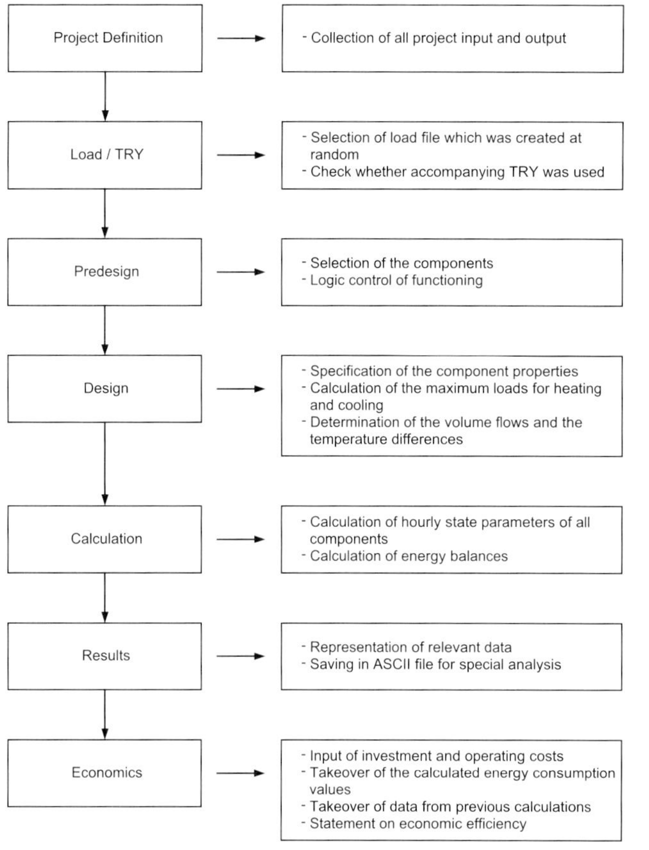

Figure 6.18
Software's structure of the TASK 25 Design tool.

Whereas a refrigeration machine can be selected directly, a description of the transmission characteristics was used for the other components. The outputs of the software include the following:

- Electrical energy demand for fans, pumps and compressors
- Energy demand of the (thermal) back-up system
- Water consumption of the cooling tower
- Water consumption when using well water
- Water consumption of the humidifiers

6.6.2 Systems

Solar-assisted air-conditioning can use a great variety of different systems, which have already been outlined in Chapters 3 and 5. Table 6.7 illustrates various sub-systems and possible configurations, which have been implemented in the software tool, for:

- Solar energy supply
- Refrigeration
- Air-conditioning

Circuit type no	1	2	3	4	5	6	7	8	9	10	11	12	13	14	15	16	17	18	19	20	21	22	23	24	25	26	27	28	29	30	31	32
Sorption regenerator				x	x	x	x				x	x	x	x				x	x	x	x				x	x	x	x	x	x	x	x
Heater for regeneration				x	x						x	x						x	x						x	x	x	x				
Air-based solar panel regeneration						x	x						x	x						x	x								x	x	x	x
Heat regenerator		x	x	x	x	x	x		x	x	x	x	x	x		x	x	x	x	x	x		x	x	x	x	x	x	x	x	x	x
Fresh air humidifier			x		x		x			x		x		x			x		x		x			x		x		x		x		x
Pre-heater								x	x	x	x	x	x	x								x	x	x	x	x	x	x	x	x	x	x
Surface cooler															x	x	x	x	x	x	x	x	x	x	x	x			x	x		
Heat pump	x	x	x	x	x	x	x	x	x	x	x	x	x	x																		
Additional air moistener								x	x	x	x	x	x	x								x	x	x	x	x	x	x	x	x	x	x
Post-heater	x	x	x	x	x	x	x	x	x	x	x	x	x	x	x	x	x	x	x	x	x	x	x	x	x	x	x	x	x	x	x	x

Table 6.7
Possible combinations of the air-handling unit components which are implemented in the simulation program.

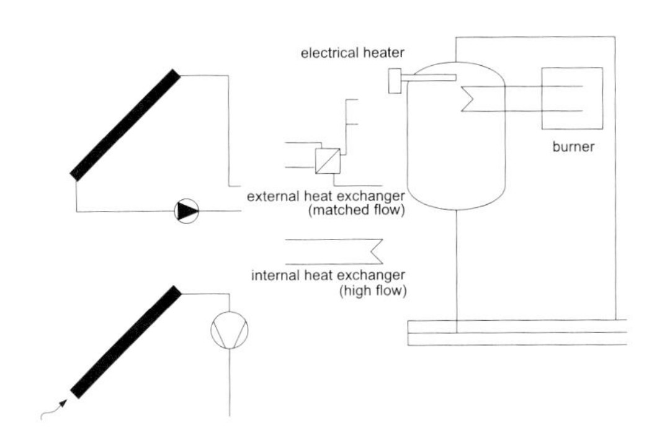

Figure 6.19
Configuration options for solar energy supply.

The Solar Sub-System

For the solar energy supply, liquid-based solar collectors or air-based solar collectors may be used. Various types of solar collectors are available in a selection diagram that already includes all the necessary coefficient/performance settings. The minimum storage capacity is calculated internally by the program in relation to the collector area. The user can define the collector area and orientation. The different configuration options for the solar energy supply are shown in Figure 6.19.

Cold Water Production Sub-System

The refrigeration module allows calculation of the following types of cold water production:

- Vapour compression machine
- Absorption system
- Adsorption system
- Free cooling by means of cooling tower
- Use of well water.

Especially for the absorption and adsorption refrigeration systems, there is a choice of different types of machines, so that the user needs to select the performance capacity and the type only. When using well water, constant water temperatures throughout the year are assumed. The free-cooling mode calculates the available cold water temperature as a function of the actual external climatic conditions defined in the meteorological data file. An overview of some of the sub-systems for providing the chilled water is given in Figure 6.20.

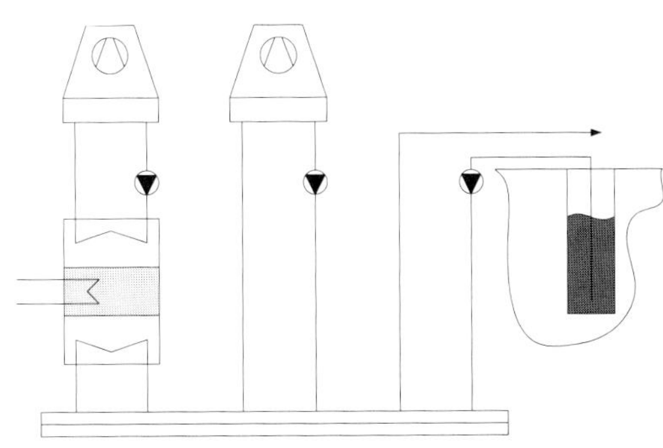

Figure 6.20
Schematic drawings of three cold water production sub-systems.

Air-Conditioning Sub-System

The various combinations within an air handling unit that are available in the software and can be investigated are listed in Table 6.7. Figure 6.21 shows the respective components; the sorptive methods (desiccant-cooling) as well as the classical methods involving surface coolers are offered.

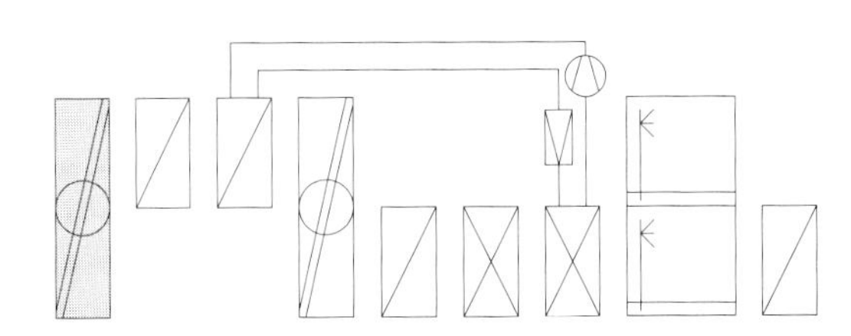

Figure 6.21
Possible configurations of air-handling units (including sorptive dehumidification).

Room Sub-System

The room components are shown in Figure 6.22. Depending on the selected systems (with/without air handling unit, with/without chiller), the conditioned space can be supplied with mechanical ventilation or infiltration only, and chilled ceiling or fan-coil systems are possible for sensible cooling. Calculations can be made for air-based systems as well as for combined water-air systems. The room components

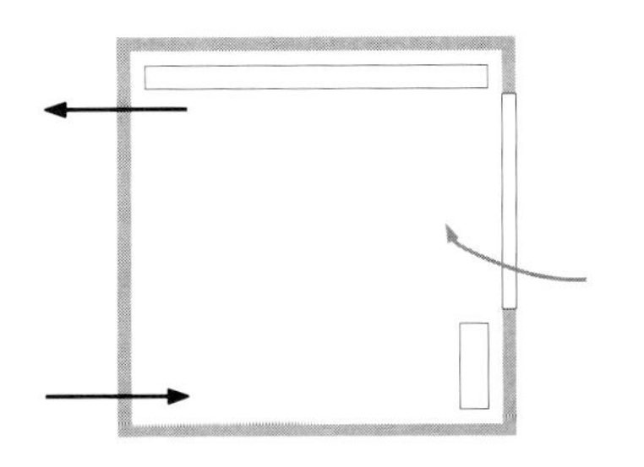

Figure 6.22
Configurations of the room components (heating, chilled ceiling/fan-coil, ventilation, infiltration).

may be presumed to be ventilator convectors or without electrical actuation (e.g., cooling blankets). Overall more than 500 different systems can be combined and simulated. The simulation tool will be commercially available in 2004.

6.7 Simulation programs

The main limitation of the design tool described above is that only calculation of pre-defined systems is possible. If another system configuration is desired in a specific project, the only way to compute a detailed energy balance is to model the system in an open simulation platform. Several computer programs for this purpose are available on the market. The most well-known in the solar energy sector is TRNSYS /6.6 /. A library which contains models for many thermal components and also for buildings enables the user to develop his own system. The following models that were developed for the IEA-SHC Task 25 design tool, have also been made available for TRNSYS:

- Single-effect absorption chiller model
- Adsorption chiller model.

Use of simulation programs like TRNSYS requires a high level of user expertise, much greater effort and a longer time to establish the model of a specific system. Other commercially available tools which contain components that are relevant to solar air-conditioning are Matlab-Simulink with the Carnot package /6.7/, ColSim /6.8/ or Smile /6.9/.

7 PERFORMANCE FIGURES

The main reason to use solar energy for air-conditioning is to reduce the consumption of conventional energy sources (i.e., fossil fuels and electricity). Therefore, energy performance is a key issue when designing solar-assisted air-conditioning systems. Of course, economic parameters are equally important and a combined analysis of energy and economic performance can help to select the configuration which leads to the highest energy savings for a given investment.

In this chapter definitions and description are given of the main performance figures which can guide a designer during the decision-making process of selecting and evaluating different systems. They are applied in specific examples and some general design considerations are presented using these parameters. Work sheets, which can be used in the design phase of a solar-assisted air-conditioning system, are provided.

7.1 Energy performance

A work sheet for assessing the energy performance of solar-assisted air-conditioning systems is presented in Table 7.1. The work sheet is developed in a general way so that all types of systems are included, independently of the type of equipment used. Only the figures which are not self-explanatory are defined next, following the same order as their presentation in Table 7.1.

The work sheet includes three main sections, which are described next.

General Data

General data: This part of the work sheet contains general data on the dimensions of the system components such as the solar collector, the buffer tank, the chiller, the air-handling unit etc.. These values have to be inserted by the user for both the reference system (R) and the solar system (S).

Results of System Design

Results of system design: These values are the results of an annual energy balance which can be derived using a simulation program, a design tool or any other calculation method. For example, the methods described in Chapter 6 can be used to obtain the values that have to be entered in this section. To allow comparison of the solar system performance with a reference system, comparable figures of a reference system have to be entered.

Energy Related Evaluation

Energy related evaluation: Energy-related performance figures are calculated, based on the annual simulation results from Section 1 of the work sheet, i.e., results of system design. The various parameters that are included in this section are defined below. The following figures are not self-explaining and described here.

The annual useful solar heat, Q_{use}, is given by

$$Q_{use} = Q_{tot} - Q_{bu} \quad (7.1)$$

where Q_{tot} is the total annual heat required for heating and cooling and Q_{bu} is the annual heat from a secondary heat source (back-up).

Energy-related comparison	Unit	Reference system (R)	Solar system (S)
0. General data			
collector type	-	-	Flat-plate collector
1 collector area (absorber)	m^2	0	200
2 volume of heat storage unit	m^3	0	18
3 volume of cold-side storage unit	m^3	0	0
4 airflow (air-handling unit)	m^3/h	0	0
5 heating power back-up heater	kW	18.8	17.9
6 nominal chiller power, compression chiller	kW	29.4	0
7 nominal chiller power, thermally driven chiller	kW	0	29.4
8 nominal power of cooling tower	kW	0	74.6
1. Results of annual energy balance for system design (e.g., obtained with simulation programs)			
9 annual total electricity consumption (including pumps, fans, control)	kWh	**24 811**	**10 997**
10 annual electricity consumption, chiller	kWh	**22 388**	**0**
11 annual required heat for cooling/dehumidification	kWh	**0**	**103 329**
12 annual required heat for heating/humidification	kWh	**14 395**	**14 395**
13 total annual heat	kWh	**14 395**	**117 724**
14 annual heat from 2nd heat source (fossil fuel)	kWh	**15 152**	**14 113**
15 annual amount of fossil heat source (primary energy)	kWh	**15 950**	**14 855**
16 annual radiation on collector	kWh	**0**	**314 980**
17 annual heat produced by solar collector	kWh	**0**	**128 803**
18 annual overall cold production (cooling, dehumidification)	kWh	**67 164**	**67 164**
19 annual cold production by compression	kWh	**67 164**	**0**
20 maximum electricity demand (maximum hourly value)	kW	**10.0**	**2.9**
21 total annual water consumption	m^3	**0**	**368.3**
2. Energy-related evaluation (computed from design results)			
22 annual useful solar heat	kWh	-	103 611
23 annual gross collector efficiency	%	-	38%
24 annual net collector efficiency	%	-	30%
25 annual COP of compression chiller	-	3.00	0.00
26 annual COP of thermally driven cold production	-	0	0.65
27 annual primary energy consumption	kWh	84 868	45 402
28 annual primary energy savings	kWh	-	39 466
29 relative primary energy savings	%	-	47%
30 specific useful net collector output	kWh/m^2	-	518
31 specific primary energy saving	kWh/m^2	-	197

Table 7.1
Energy performance work sheet; data inserted in the shaded areas are based on annual energy balances (calculated data, e.g., using one of the methods described in Chapter 6). The values entered here as an example are calculation results for a hotel in Athens in which a thermally driven chiller is used for air-conditioning in combination with an advanced flat-plate collector field. The reference system is a conventional, electrically driven compression chiller.

The annual gross collector efficiency, η_{gross}, is defined as

$$\eta_{gross} = \frac{Q_{coll}}{A_{coll} \cdot I_{coll}} , \qquad (7.2)$$

where Q_{coll} is the annual heat production of the solar system (including unused excess heat), A_{coll} is the collector (absorber) area and I_{coll} is the annual global (total) solar radiation incident on the collector plane.

The net collector efficiency, η_{net}, is defined as

$$\eta_{net} = \frac{Q_{use}}{A_{coll} \cdot I_{coll}} . \qquad (7.3)$$

The net collector efficiency can be compared to the annual gross collector efficiency. The difference indicates the amount of solar energy that cannot be used because there is no demand for the solar heat.

The annual primary energy consumption, E_{PE} is defined as

KEY FIGURE/ PRIMARY ENERGY BALANCE

$$E_{PE} = \frac{Q_{bu}}{\varepsilon_{fossil}} + \frac{E_{elec}}{\varepsilon_{elec}} \qquad (7.4)$$

where ε_{fossil} is the primary energy conversion factor of the fossil fuel used for the back-up heat source (e.g., gas, oil), E_{elec} is the total electricity consumption of the solar driven system (i.e., for electrically driven components, for example pumps) and ε_{elec} is the primary energy conversion factor for electricity production.

The annual primary energy savings, $E_{PE,save}$, is given as

$$E_{PE,\,save} = E_{PE,\,reference} - E_{PE,\,solar} \qquad (7.5)$$

,

where $E_{PE,reference}$ is the annual primary energy consumption of the reference system and $E_{PE,solar}$ is the annual primary energy consumption of the solar driven system.

The relative primary energy savings $E_{PE,save,rel}$ are defined as

$$E_{PE,\,save,\,rel} = \frac{E_{PE,\,save}}{E_{PE,\,reference}} \qquad (7.6)$$

and the primary energy savings per unit area of the solar collector ,$E_{PE,save,spec}$, are defined as

$$E_{PE,\,save,\,spec} = \frac{E_{PE,\,save}}{A_{coll}} \qquad (7.7)$$

.

The primary energy savings per unit area of solar collector,$E_{PE,save,spec}$, give an idication of the contribution of each square meter of collector field to the energy saving of the entire system.

In Chapter 6, the solar fraction was used as a performance index and is defined as the fraction of required heat for a certain application which is delivered by the solar system. However, this parameter is difficult to judge in some cases, since it does not reflect the full picture of the energy balance. Particularly for solar cooling systems, in which different energy sources may serve as a back-up, it may be difficult to define the solar fraction appropriately. Therefore, since estimation of the primary energy savings is the main goal, it is recommended to use the corresponding parameter to quantify the energy performance of a solar-assisted air-conditioning system.

For complete system performance assessment, it is necessary to consider the energy consumption for the entire year. Therefore, in all performance figures described above, the energy used for heating is also taken into account. Especially in regions with high solar radiation availability during winter (heating season), the solar system can also cover a high percentage of the heating energy demand (see for instance Section 6.4.3).

7.1.1 Primary energy balance

In the previous section different parameters were defined to compare solar-assisted air-conditioning systems with conventional reference systems. The key parameter to measure energy-related performance is the primary energy saving due to the solar-assisted system.

A very general analysis of the primary energy consumption of a solar-assisted system can be made by comparing the design values of a solar-thermally driven system with those of a conventional system. The specific primary energy consumption of a conventional chiller, driven by electricity, $PE_{spec,conv}$, can be calculated as follows:

$$PE_{spec,\,conv} = \frac{1}{\varepsilon_{elec} \cdot COP_{conv}} \tag{7.8}$$

where COP_{conv} denotes the COP of the conventional chiller. $PE_{spec,conv}$ is expressed in kWh of primary energy per kWh of cold production.

It is assumed that the COP takes into account the average electricity consumption including the one needed for heat rejection at the condenser, as is valid for most compact chillers in the low power range, e.g., below 100 kW, which do not use a separate wet cooling tower.

For a solar-thermally driven chiller, which uses a fossil-fuelled back-up heater as a secondary heat source, the specific primary energy consumption, expressed in kWh of primary energy per kWh of cold production, is given by:

$$PE_{spec,\,sol} = \frac{1}{\varepsilon_{fossil} \cdot COP_{thermal}} \cdot (1 - SF_{coll}) + PE_{spec,\,cooling\ tower} \tag{7.9}$$

where SF_{cool} is the solar fraction for cooling, $COP_{thermal}$ is the COP of the thermally driven chiller and $PE_{spec,cooling\ tower}$ is the specific primary energy consumption of the cooling tower per unit of cold produced by the chiller (kWh_{PE}/kWh_{cold}).

From a simple energy balance for the thermally driven chiller, the specific primary energy consumption of the cooling tower per unit of cold production, $PE_{spec,cooling\ tower}$, including the electricity needed for the circulation pump of the cooling water cycle, can be expressed as follows:

$$PE_{spec,\,cooling\ tower} = \frac{E_{spec,\,cooling\ tower}}{\varepsilon_{elec}} \cdot \left(1 + \frac{1}{COP_{thermal}}\right) \tag{7.10}$$

In this equation $E_{spec,cooling\ tower}$ is the specific electricity demand of the cooling tower per unit of cooling energy (heat rejection) including the circulation pump of the cooling water cycle and it is expressed in $kWh_{el}/kWh_{cooling}$; respective values were given in Section 3.3.1.

Comparison Example

The primary energy consumption of a cooling tower is presented in Figure 7.1 as function of its electricity consumption and the $COP_{thermal}$, calculated according to Equation 7.10. For example, a $PE_{spec,cooling\ tower}$ of 0.337 kWh_{PE}/kWh_{cold} results for a specific electricity consumption of 0.05 $kWh_{el}/kWh_{cooling}$ and a $COP_{thermal}$ of 0.7.

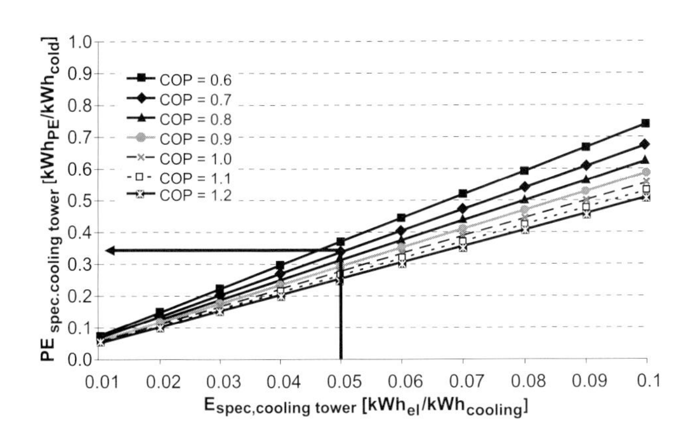

Figure 7.1
Primary energy consumption of the cooling tower per kWh of produced cold as a function of the specific electricity consumption of the cooling tower per kWh of cooling energy (heat rejection) for different values of the COP of the thermally driven chiller.

PRIMARY ENERGY PERFORMANCE

Figures 7.2 to 7.4 present the primary energy consumption of a solar-assisted thermally driven cooling system with a fossil-fuelled heat back-up as a function of the solar fraction for cooling (i.e., the fraction of the heat needed for the thermally driven chiller that is produced by solar energy). A solar fraction with a value of zero means that all the heat is delivered from the fossil-fuelled back-up heater.

Figure 7.2 is valid for a specific electricity consumption of the cooling tower system of 0.02 kWh of electricity per kWh of cooling (heat rejection), Figure 7.3 is valid for 0.05 $kWh_{el}/kWh_{cooling}$ and Figure 7.4 is valid for 0.08 $kWh_{el}/kWh_{cooling}$.

All figures above are valid for a primary energy conversion factor of 0.36 (kWh of electricity per kWh of primary energy) and a primary energy conversion factor for heat from fossil fuels of 0.9 (kWh of heat per kWh of primary energy).

Figures 7.2 to 7.4 also show the primary energy consumption of a conventional, electrically driven compression chiller. The corresponding specific primary energy consumption, $PE_{spec,conv}$, is calculated using Equation 7.8. The upper line in the figures is valid for a compression system with an overall COP of 2.5, which typically characterises a system with quite a low cooling power. The lower line is valid for a very efficient system with a COP of 4.5.

As example assuming a $COP_{thermal}$ of 0.7, a $E_{spec,cooling\ tower}$ equal to 0.05 and a solar fraction for cooling of 0.7 the specific primary energy consumption is 0.81 kWh of primary energy per kWh of cold production (see arrows in Figure 7.1 and Figure 7.3). The specific primary energy consumption for the example is about 27 % lower than that of a conventional compression chiller with an effective COP of 2.5, but it is about 32 % higher than that of a conventional chiller with an effective COP of 4.5.

The analysis shows that it is necessary to reach a certain value of the solar fraction in order that a solar-assisted cooling system achieves a lower primary energy consumption than a conventional system using an electrically driven compression chiller. The system performance with regard to primary energy improves when the COP of the thermally driven chiller increases, the solar fraction increases and the specific electricity consumption of the cooling tower system decreases.

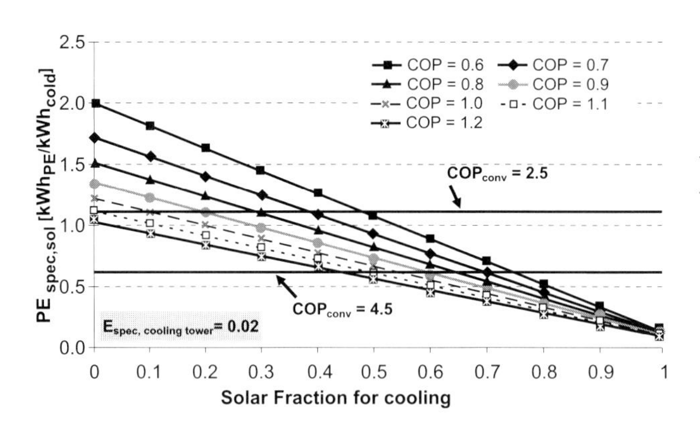

Figure 7.2
Primary energy consumption of a solar-assisted cooling system with fossil-fuelled thermal back-up as a function of the solar fraction for cooling for different values of the COP of the thermally driven cooling system; the figure is valid for a specific electricity consumption of 0.02 kWh_{el} per kWh of cooling (heat rejection), a primary energy conversion factor for electricity of 0.36 and a primary energy conversion factor for heat from fossil fuels of 0.9.

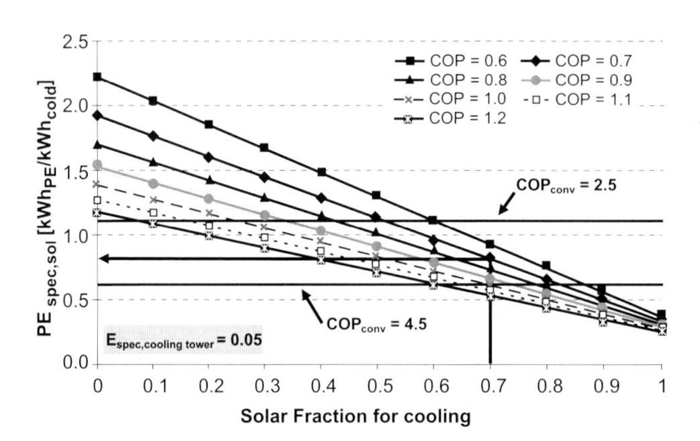

Figure 7.3
Diagram similar to Figure 7.2; the figure is valid for a specific electricity consumption of 0.05 kWh_{el} per kWh of cooling (heat rejection).

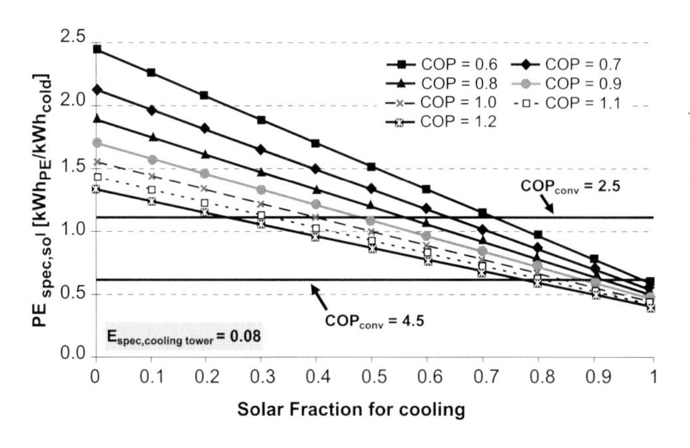

Figure 7.4
Diagram similar to Figure 7.2; the figure is valid for a specific electricity consumption of 0.08 kWh_{el} per kWh of cooling (heat rejection).

Desiccant Cooling Systems

The figures defined above can be used in a similar way for desiccant cooling air-handling units, which have to be compared to conventional air-handling units that use a compression chiller for both sensible cooling and dehumidification (i.e., cooling the air below the dew-point). Desiccant systems with rotors do not use a cooling tower. Therefore, electricity consumption of a cooling tower has not to be taken into account. However, due to the additional components in the air-handling unit, mainly the desiccant rotor, a higher pressure drop occurs compared to conventional air-handling units, which leads to increased electricity consumption for the fans compared to a similar conventional system. Therefore, the primary energy consumption of the cooling tower, $PE_{spec,cooling\ tower}$ in Equation 7.9 has to be replaced by the corresponding increased primary energy consumption that is due to the higher electricity consumption of the fans.

The consequences of this analysis for the design of solar-assisted air-conditioning are:

1) Using a thermally driven cooling or air-conditioning system with a comparatively low COP and a fossil-fuelled heat source employed as the only back-up, requires a high solar fraction in order to achieve significant primary energy savings. The goal of the design, with regard to energy savings, is to guarantee a high solar fraction by using a sufficiently large solar collector area, integrating sufficient heat storage capacity and possibly using other storage possibilities, such as the building thermal mass (e.g., slab cooling) to maximise the use of solar heat.

2) An alternative is a system which uses a conventional chiller as a back-up system. In this case, each unit of cooling energy provided by the solar-thermally driven system reduces the cooling power to be delivered by the conventional unit. Such a design allows some primary energy savings even for low solar fractions. In such a design, the solar system serves mainly to reduce the electricity consumption.

3) Solar-thermally autonomous systems do not use any other source for cold production and therefore always work at the upper limit with a solar fraction of 100 %.

4) Systems employing a thermally driven chiller with a high COP may be designed with a smaller solar fraction even if a fossil-fuelled secondary heat source is used. The reason is that the heat from the fossil fuel burner is also converted with a higher COP, which is competitive to a conventional system in terms of primary energy.

5) In all cases, use of the solar collector should be maximised by also supplying heat to other loads such as the heating system or hot water production.

Although the above analysis is rough and neglects many details such as part load behaviour and solar contributions during the heating period, it demonstrates the general difficulties encountered during the design of the solar sub-system for a solar-assisted air-conditioning system.

Therefore, it is absolutely essential to calculate a detailed energy balance during the design phase of a project, in order to choose the configuration that maximises the primary energy savings.

7.2 Economic performance

The second important aspect, which has to be considered in reaching a good decision about the installation of a solar-assisted air-conditioning system, is the cost performance. First, it is necessary to sum up all the investment (initial) costs for each component of the entire system, and its installation. These values are listed in the first part of the work sheet shown in Table 7.2. The sum of the corresponding values results in the total investment cost for the reference system and the solar system. In some countries the capital cost of the solar sub-system can be reduced since subsidy programmes are available; the latter should be taken into account when calculating the economical performance of the solar-driven air-conditioning system.

The second part of the work sheet is used to calculate the annual costs. An annuity factor is used in order to calculate the annual capital cost for the initial investment for both the conventional equipment and the solar system. Different values for the solar part of the system and the chiller or the air-handling unit can be used, since for instance different interest rates or equipment lifetimes have to be considered. To calculate the cost of energy and water, the annual consumption figures from the energy performance work sheet are used together with prices for electricity, fossil fuel and water. The total annual cost is calculated by summing up the capital cost, maintenance and inspection cost, energy cost and water cost. The following values are defined in order to calculate cost performance figures:

COST PERFORMANCE FIGURES

The annual extra cost of the solar system, $\Delta C_{annual,sol}$, is given by:

$$\Delta C_{annual,\ sol} = C_{annual,\ sol} - C_{annual,\ ref} \qquad (7.11)$$

where $C_{annual,sol}$ is the annual cost for the solar variant and $\Delta C_{annual,ref}$ is the annual cost for the reference system.

If the annual cost for operation and maintenance of the solar system is lower than for the reference system due to the lower energy consumption and related cost, this cost saving, $C_{oper,annual,sol}$, is given by:

$$\Delta C_{oper,\ annual,\ sol} = C_{oper,\ annual,\ sol} - C_{oper,\ annual,\ ref} \qquad (7.12)$$

where $C_{oper,annual,sol}$ is the annual operation and maintenance cost for the solar variant and $C_{oper,annual,ref}$ is the annual operation and maintenance cost for the reference system.

PAYBACK TIME

Based on these definitions, two parameters are defined which can be used to compare the cost performance of different solar system designs, in which for instance different collector areas, collector types or storage dimensions are used. The first parameter is the payback time, $\tau_{payback}$, which is given by:

$$\tau_{payback} = \frac{C_{invest,\ tot,\ sol} - C_{invest,\ tot,\ ref}}{\Delta C_{oper,\ annual,\ sol}} \qquad (7.13)$$

where $C_{invest,tot,sol}$ is the total investment cost for the solar system and $C_{invest,tot,ref}$ is the total investment cost for the reference system.

COST OF SAVED PRIMARY ENERGY

Of course, the definition of payback time delivers a sensible result only, if the annual operation cost of the solar system is lower than that of the reference system. The second performance parameter, which combines energy and cost, is the cost of saved primary energy, $C_{PE,saved}$, which is given by

$$C_{PE,\ saved} = \frac{\Delta C_{annual,\ sol}}{E_{PE,\ saved}} \quad [€/kWh_{PE}] \qquad (7.14)$$

where $E_{PE,saved}$ is the annual primary energy saving of the solar system compared to the reference system.

Similarly, this parameter makes sense only if the total annual cost of the solar system is higher than that of the reference system. This will be probably valid in most cases under present cost and price conditions. In this case, this parameter can be used to compare different measures to save primary energy, since it indicates the actual cost for saving one unit of primary energy when comparing a solar-assisted system with a reference system. For example, using the $C_{PE,saved}$, a solar-assisted air-conditioning system can be compared with other energy-saving measures, e.g., methods of improved building construction which reduce the demand for air-conditioning or the installation of advanced, more expensive conventional systems which do not use solar energy. And this parameter allows also the comparison among different solar collector field options e.g., configuration with different collector types or different sizes of the collector field.

Economic assessment	Unit	Reference system (R)	Solar system (S)
1. Investment costs			
32 solar collector system including supporting structure	EURO	0	50 000
33 heat storage unit	EURO	0	9 000
34 (additional) heat source (e.g. gas burner)	EURO	7 520	7 160
35 installation costs (including piping, pumps, ...)	EURO	10 000	10 000
36 air-handling unit or desiccant air-handling unit	EURO	0	0
37 compression chiller	EURO	11 760	0
38 thermally driven chiller	EURO	0	11 760
39 cooling tower	EURO	-	2 611
40 cold storage unit	EURO	0	0
41 pumps	EURO	750	750
42 control system	EURO	8 000	8 000
43 planning costs	EURO	3 803	9 928
44 total investment cost without funding subsidies	EURO	41 833	109 209
45 funding (investment support)	EURO	0	0
46 funding related to solar collector	EURO	0	25 000
47 final total investment cost	**EURO**	41 833	84 209
2. Annual costs			
48 annuity factor, conventional equipment	%	6.7	6.7
49 annuity factor, solar system (collector, storage)	%	-	5.1
50 capital cost	EURO	2 811	4 675
51 cost for maintenance, inspection	EURO	837	1 594
52 annual electricity cost (consumption)	EURO	4 962	2 199
53 annual electricity cost (peak)	EURO	1 000	290
54 annual heat cost (fossil fuel)	EURO	606	565
55 annual water cost	EURO	0	1 105
56 total annual cost	**EURO**	10 216	10 428
57 annual extra cost of solar system	EURO	-	212
58 annual operation and maintenance cost	EURO	7 405	5 753
3. Comparative evaluation			
59 payback time	a	-	25.7
60 cost of saved primary energy	EURO/kWh	-	0.005

Table 7.2
Performance work sheet; values inserted in the shaded areas are specific to each project.

7.2.1 Energy-economic design study - example

In order to demonstrate the use of the various parameters and the work sheets described above, an example is presented, in which the SACE method (see Section 6.5) was used to compare annual energy balances for both, solar variant and reference system. The goal in this case example is to design a solar-assisted chilled-water system, which provides cooling power to air-condition an office building in Madrid. The meteorological and load data presented in Chapter 6 are used. The reference case employs a conventional electrically driven compression system for cooling and a gas burner for heating (scheme according to System 10 in Figure 6.17) . Several scenarios for different solar-assisted system configurations will then be considered and compared to the reference case (conventional design).

Basis Case Definition

The base-case system (configuration 1) is assumed to consist of a stationary CPC collector and a back-up gas heater, which provide heat to an absorption chiller (scheme according to System 2 in Figure 6.17) . All the necessary information and detailed design data are summarised in Appendix 2. As a first step, for the base case (stationary CPC collector, absorption chiller, thermal back-up), the annual performance for different dimensions of the key components in the solar sub-system is studied, including the:

- Solar collector area, ranging between 100 m² and 280 m² and the
- Heat buffer tank volume, ranging between 0.03 m³ per m² of collector area to 0.18 m³ per m² of collector area.

The annual performance for all the different combinations of the collector area and the storage dimensions is presented in Figures 7.5 and 7.6. The primary energy savings for the studied system combinations of the base case are presented in Figure 7.5 and the cost of the saved primary energy, as defined in Equation 7.14, is presented in Figure 7.6. All combinations lead to primary energy savings and all of them lead to higher annual costs compared to the reference case. The lowest value for the cost of saved primary energy, $C_{PE,saved}$, for this particular example is achieved with a collector area of 130 m² and a storage volume of about 23 m³. The cost per kWh of primary energy saving in this case is 10.6 €-cent.

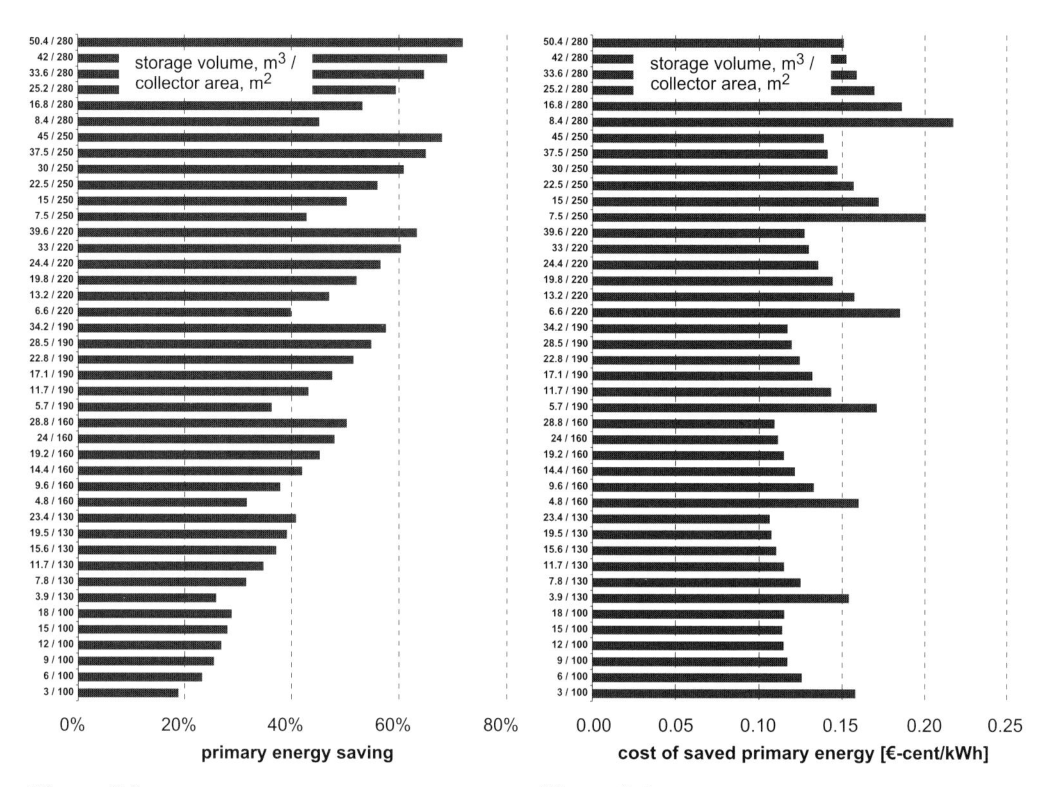

Figure 7.5
Primary energy savings for the studied system configurations of the base case.

Figure 7.6
Cost of saved primary energy for the studied system configurations of the base case.

Note: the values defining each bar denote the storage size, m³, and the collector area, m².For instance: 9.4 / 160 stands for a system with a collector area of 160 m² and a hot water storage volume of 9.4 m³.

PARAMETRIC STUDY

Different system configurations are defined next by changing some of the components with respect to the base case (configuration 1: stationary CPC/absorption chiller/thermal back-up) system as follows:

Configuration 2: Use of an electrically driven chiller as a back-up instead of the gas burner;

Configuration 3: Use of an adsorption system instead of the absorption chiller and a thermal back-up gas burner;

Configuration 4: Use of an adsorption system instead of the absorption chiller and an electric back-up (electrically driven chiller);

Configuration 5: Use of a selective flat-plate collector instead of the stationary CPC collector, an absorption chiller and a thermal back-up (gas burner);

Configuration 6:	Use of a selective flat-plate collector instead of the stationary CPC collector, an absorption chiller and an electric back-up (electrically driven chiller);
Configuration 7:	Use of a selective flat-plate collector instead of the stationary CPC collector, an adsorption chiller and a thermal back-up;
Configuration 8:	Use of a selective flat-plate collector instead of the stationary CPC collector, an adsorption chiller and an electric back-up.

The main system characteristics for all the configurations are summarised in Table 7.3. The table also includes the annual cost, the primary energy savings and the cost of saved primary energy. For each configuration, the same parametric study on the size of the solar collector and the heat storage unit was carried out, to select the combination that leads to the lowest cost of saved primary energy; the corresponding dimensions are also listed in Table 7.3. A graphical presentation of the major performance parameters is shown in Figure 7.7.

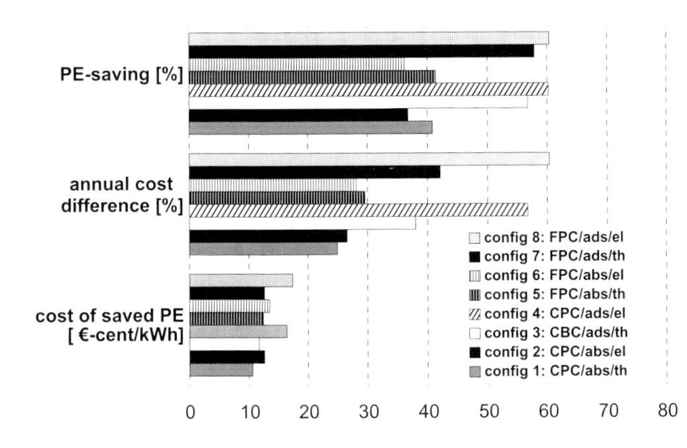

Figure 7.7
Results of the comparison for the different system configurations (example: Office building in Madrid).

RESULTS OF PARAMETRIC STUDY

The following results can be drawn from the comparative analysis:

- The annual cost of all solar configurations is higher than that of the conventional reference system. The difference between the annual cost of solar and conventional systems is lower if an absorption chiller is used than for an adsorption chiller; this is mainly due to the high initial cost of adsorption chillers. The lowest value in annual cost among the solar variants studied is found for the base-case system (absorption chiller with stationary CPC collector and thermal back-up).
- The collector area has to be significantly larger if flat-plate collectors are applied; the increase is 50 - 90 % compared to a stationary CPC collector depending on the system considered.
- The primary energy saving is higher for the configurations with adsorption chillers. The reason is that these machines operate with lower driving temperatures, thus increasing the useful collector output.
- The lowest cost of saved primary energy is also achieved with the base-case configuration. This cost is about 10 €-cent per kWh.
- The system design leading to the lowest cost for saved primary energy is quite different for systems using a thermal back-up and for those using a back-up on the cold production side. In general, the use of a conventional, electrically driven chiller as the back-up leads to smaller collector areas and storage volumes than the use of a back-up heat source, if the minimum of cost of saved primary energy is taken as goal function.

Influence of Subsidies

The financial incentives were evaluated next, in order to compare the various configurations with the conventional system in economic terms, e.g., to determine the subsidies for which the annual cost falls below the annual cost of the conventional (reference) system. The subsidy values were assumed in terms of investment funds per unit collector area, as they are offered in several countries. The calculations are based on the base case according to configuration 1 in Table 7.3.

configuration No: abbreviation	solar collector type	collector area	collector area	storage tank volume	chiller type	type of back-up	total annual cost	primary energy saving		cost of saved primary energy
	-	m^2	$m^2/kW_{chiller}$	m^3	-	-	€	MWh	%	€-cent/kWh
reference	-	-	-	-	compression	-	11 100	-	-	-
config 1: CPC/abs/th	stationary CPC	130	2.77	23	absorption	thermal	13 900	26.1	40.9	10.6
config 2: CPC/abs/el	stationary CPC	70	1.49	8	absorption	electric	14 100	23.5	36.8	12.6
config 3: CBC/ads/th	stationary CPC	130	2.77	20	adsorption	thermal	15 300	36.3	56.8	11.7
config 4: CPC/ads/el	stationary CPC	130	2.77	16	adsorption	electric	17 400	38.5	60.2	16.4
config 5: FPC/abs/th	sel. Flat-plate	250	5.33	45	absorption	thermal	14 400	26.5	41.4	12.4
config 6: FPC/abs/el	sel. Flat-plate	100	2.13	9	absorption	electric	14 300	23.2	36.2	13.5
config 7: FPC/ads/th	sel. Flat-plate	190	4.05	23	adsorption	thermal	15 800	37.0	57.9	12.6
config 8: FPC/ads/el	sel. Flat-plate	190	4.05	17	adsorption	electric	17 800	38.6	60.4	17.4

Table 7.3

Comparative results for the studied system configurations. For each configuration (1-8) the system combination (collector size, storage volume) with the lowest value of the cost of saved primary energy was selected.

The results of these calculations are shown in Figure 7.8. It turns out that a subsidy in the range of 75 % of the collector cost (collector including support structure without storage, pumps, controls etc.) leads to the same annual cost as the conventional reference system, in other words, a cost break-even is achieved. For the initial cost of the entire system, including chiller, storage, installation and planning cost etc., this would correspond to a subsidy in the range of 27 %. With higher subsidies, the annual cost performance of the solar configurations is positive and primary energy savings also lead to cost savings.

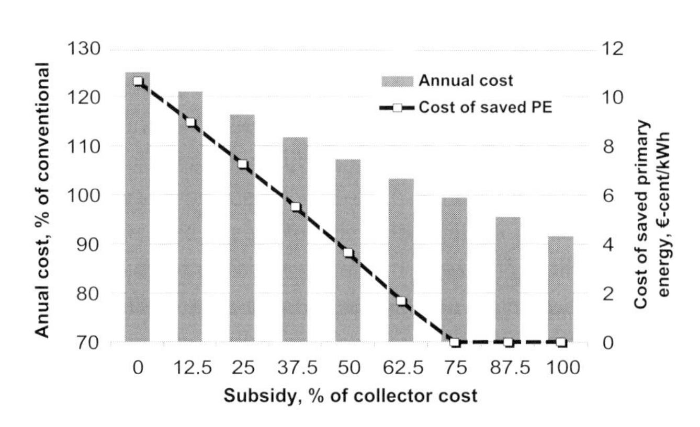

Figure 7.8

Annual cost of solar system configurations, expressed as a percentage of the corresponding cost of the conventional reference system, and cost of saved primary energy versus subsidy, expressed as a percentage of collector cost.

Summary

The objective of this example is to demonstrate the use of the performance parameters to compare different system configurations with regard to their potential energy savings and cost performance. The cost of saved primary energy is a very useful figure of merit to compare different energy saving measures. Although this was only one example, the results of this economic analysis are typical for the present cost situation of solar-assisted air-conditioning systems. Without financial incentives, solar-assisted air-conditioning cannot, for the time being - given the energy prices and system component costs that were used - be economically competitive to conventional systems. However, with funds that support energy-saving measures and the use of systems that exploit renewable energy sources - as are available in several countries - solar-assisted air-conditioning can become financially competitive, and can be considered a promising approach for the future.

7.3 Environmental benefits

CO_2 SAVINGS

The main motivation to apply renewable energy is to replace systems that use conventional energy sources with systems that have a lower environmental impact. Any primary energy savings result in CO_2 reduction. However, the situation for solar-assisted air-conditioning systems becomes more complex if a thermally driven back-up is used. As was shown earlier, considerable care has to be taken to design solar-assisted systems that ensure primary energy savings. The total CO_2 savings are estimated by taking into account the CO_2 emission due to conventional electricity generation, the production of fossil fuels for heating, etc.. How much CO_2 is saved by a certain installation due to reduction of the primary energy consumption depends on the composition of the energy sources used, i.e., the CO_2 emission rates for production of conventional electricity and the CO_2 emission rates for the production of fossil fuels.

Environmental issues	Unit	Solar System (S)
61 saved electric energy	kWh	13 814
62 CO_2 saving due to electricity saving	kg	9 670
63 saved electric power	kW	7.1
64 saved fossil fuel energy for heat	kWh	1 095
65 CO_2 saving due to heat saving	kWh	306
66 water saving	m^3	-368.3
67 overall primary energy saving	kWh	39 466
68 total CO_2 saving	**kg**	**9976.1**
69 material pair solar system (refrigerant/sorbent)	-	water / silica gel
70 refrigerant reference system	-	FKW 134a
71 environmental advantage of solar system	-	environmentally friendly refigerant (no greenhouse potential)

Table 7.4
Environmental issues work sheet; the calculated values depend on the specific CO_2 emission rates assumed for the production of electricity and fossil fuels for heating.

A work sheet to assess environmental issues is shown in Table 7.4. The CO_2 emission calculated in the work sheet depends on the assumed specific emission rates of CO_2 for the production of electricity and fossil fuels for heating. In addition to the CO_2 savings, the advantage of using environmentally friendly refrigerants is highlighted, which do not have any ozone-depleting or global-warming potential. The conventional (reference) system is assumed to use R-134a as refrigerant in the example shown here.

ENVIRONMENTALLY FRIENDLY REFRIGERANTS

In order to complete the picture about the environmental performance of solar-assisted air-conditioning systems, a complete life-cycle analysis should be carried out and compared to a corresponding analysis of conventional systems. However, this is a complex task which is beyond the scope of this handbook.

8 DESIGN EXAMPLES

Three design examples are presented in this chapter. The first project is the largest European solar-cooling installation at a cosmetics factory in Greece. The second example is extracted from a project currently carried out in Guadeloupe. The plant provides cooling for an office building. The third is an existing system in Germany, which provides conditioned air to a seminar room.

8.1 The SARANTIS cosmetics factory at Inofita Viotias, Greece

GENERAL DESCRIPTION

This is the largest European solar-cooling installation, and is located at a cosmetics factory in Greece. The facility includes a total area of 2 700 m² flat-plate solar collectors, coupled with two adsorption chillers (total cooling power 700 kW) for meeting 40 % of the total cooling load of the factory (22 000 m²). The installation was completed in 1999.

Main information about the installation:

Owner:	SARANTIS S.A., Greece
Contractor:	SOLE S.A., Greece (Design, installation, monitoring, maintenance)
Solar Collectors:	SOLE S.A., Greece
Cooling machines:	GBU mbH (adsorption units)

8.1.1 Building and load

Figure 8.1
General view of the Sarantis S.A. cosmetics factory.

The factory is located in an industrial zone (Inofita Viotias) about 50 km north-east of Athens. The total air-conditioned area is 22 000 m² and 130 000 m³. The facility operates 5 days a week.

Geographical co-ordinates:
- Latitude: 38° 23' N
- Longitude: 23° 06' E

Daily mean global horizontal radiation:
- Summer: 6.64 kWh/m²day
- Winter: 2.35 kWh/m²day
- Average: 4.39 kWh/m²day

Mean outdoor temperature (°C) and relative humidity (%):
- Summer: 26°C and 55 %
- Winter: 8°C and 77 %
- Average: 17.4°C and 66 %

The peak cooling load is estimated to be 1750 kW. Dehumidification is not a major concern, given the local weather conditions.

8.1.2 Air-conditioning concept and design of equipment

The installation was designed with a good balance between solar cooling (40 %) and the conventional system (60 %). The general system diagram is shown in Figure 8.2. The operation of the entire installation is fully automated. A PLC system transfers the information from the site to the SOLE offices for monitoring and control of operating conditions.

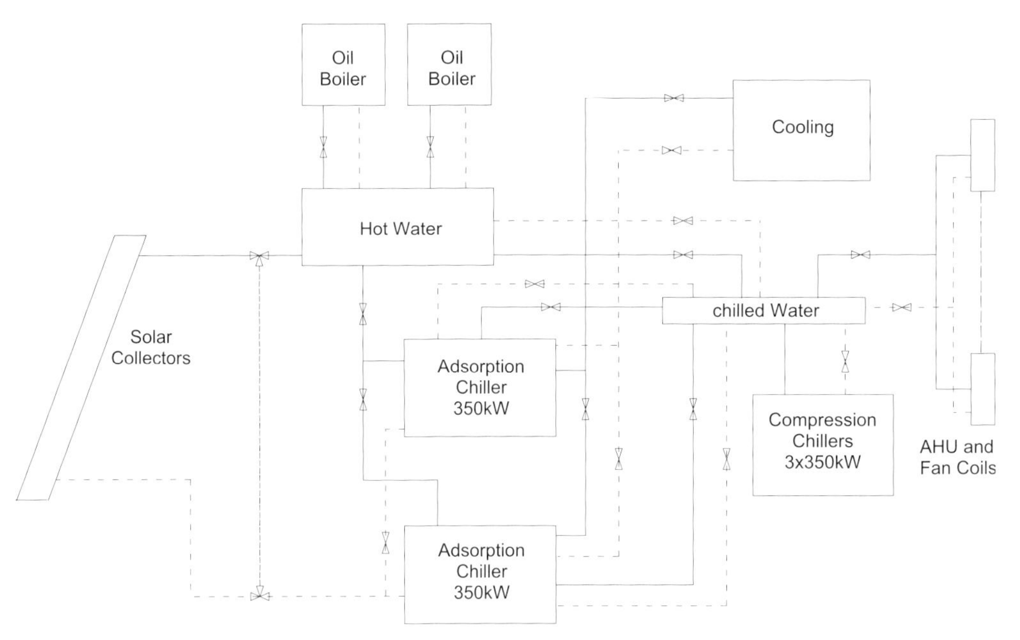

Figure 8.2
General system diagram.

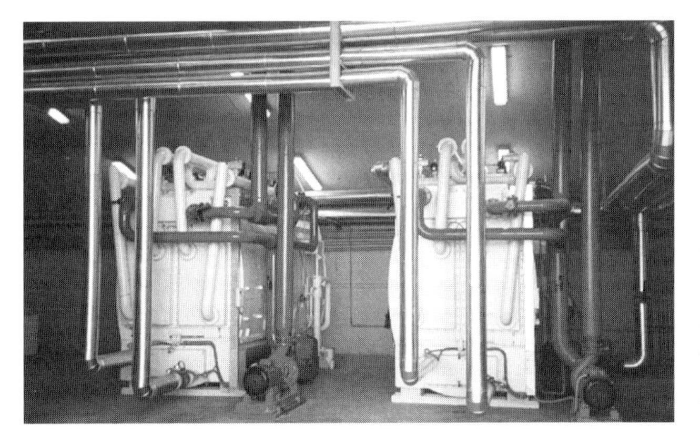

Figure 8.3
Adsorption chillers.

The installation includes two adsorption chillers (Figure 8.3), each 350 kW. Adsorption chillers were selected since they require a lower supply temperature (around 70°C, 120 m^3/h) from the solar system, compared to absorption chillers. The annual electricity consumption averages 6500 kWh/year. The annual water consumption is about 2500 m^3, averaging 15 - 18 m^3/day.

Figure 8.4
Back-up heating system.

During summer, the chilled water is supplied at 7°C to 12°C (depending on the load) and 240 m^3/h to the indoor air-handling units (AHU). The AHU heat exchangers have been over-dimensioned to facilitate their operation at the relatively high chilled-water supply temperature. The total air supply flowrate to the building is 325 000 m^3/h, with a supply air temperature of 22°C to 26°C and 50 % relative humidity. The chilled water storage volume is 60 m^3. The back-up cooling system consists of three compression chillers with a capacity (cooling power) of 350 kW each.

Two oil-fired boilers (1200 kW each) are used as a back-up heating system (Figure 8.4). During winter, the solar collectors supply water at 55°C directly to the heated spaces. The solar cooling facility is maintained by the design/installation team (SOLE S.A.) under subcontract to the owner. No major problems have been encountered.

8.1.3 Collector and solar system design

Figure 8.5
2 700 m^2 flat-plate solar collector array.

The solar collector array has a total absorber area of 2 700 m^2 (Figure 8.5) and consists of flat-plate selective-absorber solar collectors (Climasol 350), manufactured by SOLE S.A. The average solar collector efficiency is 67 %.

The solar collector medium is water, without any additives. The hot water supply temperature to the adsorption chillers is around 70°C, 120 m^3/h. The installation has a small solar buffer tank of 6 m^3.

8.1.4 Energy performance

According to monitoring data, the solar assisted cooling system covers about 45 % of the cooling load for the period April-October. The delivered cooling energy is about 780 000 kWh/year. The delivered thermal (heating) energy in winter is 900 000 kWh/year. The delivered gross energy from the collectors is about 2 200 000 kWh/year.

8.1.5 Economics

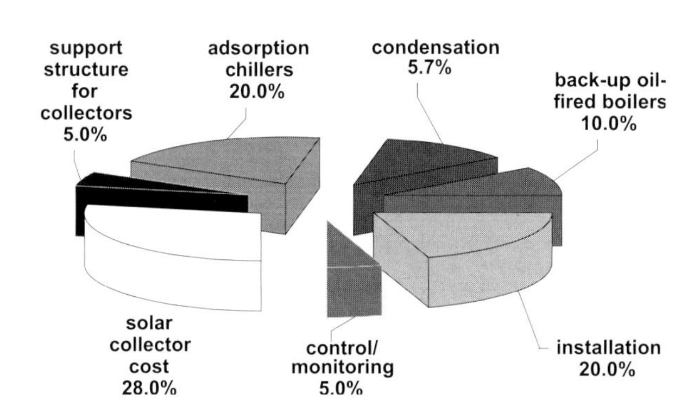

Figure 8.6
First costs breakdown for the entire system.

The total initial investment for the entire installation, including the solar and the back-up cooling/heating systems (but not including the air-handling units and piping within the building) was 1 400 000 €, without value added tax (VAT). The project was subsidised by 50 % from the national 'Operational Programme for Energy'.

The breakdown of the total investment cost is illustrated in Figure 8.6. The cost for the solar collectors was 150 €/m^2 including piping and installation. Together with the steel support structure of the solar collectors array this represented 33 % of the total cost.

8.2 An office building in Guadeloupe

Figure 8.7
Main entrance, north-west facade of the office building in Guadeloupe.

This office has been built in Guadeloupe, at Basse Terre. This building, of high environmental quality, includes a cooling system that uses chilled water produced from solar energy as a demonstration project. The project was initiated in 1998 and the offices are in operation since 2003.

8.2.1 Building and load

GENERAL DESCRIPTION

This project concerns a two-storey office building, build in length and slightly curved (Figures 8.7 to 8.10). The main axis runs North-East/South-West. The total floor area is 1000 m^2, but only 570 m^2 are cooled by the solar system, i.e., 36 office rooms.

Figure 8.8
Bird's eye view of the entire building.

The basic data used for the system dimensioning are as follows:

- Latitude: 16.3 ° N
- Altitude: 40 m
- Design air temperature, summer: 32°C
- Average wet-bulb temperature in summer: 27°C
- Daily difference: 7 K
- Average dry-bulb temperature in winter: 20°C

Figure 8.9
View of the north facade.

Figure 8.10
Installation of the collectors on the roof.

Month	Average daily ambient air temperature °C	Average daily global horizontal radiation total Wh/m².d
Jan	24.4	4435
Feb	24.5	4844
Mar	24.9	5142
Apr	25.9	5431
May	26.9	5667
Jun	27.5	5532
Jul	27.6	5497
Aug	27.6	5339
Sep	27.4	5124
Oct	26.9	4557
Nov	26.2	4153
Dec	25.2	3861

Table 8.1
Average monthly data (station Le Raizet, Guadaloupe).

The average monthly data for the ambient air temperature and global horizontal radiation are given in Table 8.1.

The desired indoor conditions are as follows:
- Dry-bulb temperature: 26°C
- Relative humidity: 50 %

The air-conditioning system is operated during office hours, that are from 7 a.m. to 5 p.m., from Monday to Friday.
Ventilation: The total fresh air supply rate is 1470 m³/h.

The maximum cooling load is 79 kW with:
- 57 kW without ventilation
- 22 kW due to the fresh air intake.

8.2.2 Air-conditioning concept and design of equipment

The cold distribution network consists of a chilled-water circuit and is produced according to the following principle (see Figure 8.11):

- Pre-cooling of the chilled water loop (nominal temperatures: 7 - 12°C) by an absorption chiller (30 kW), driven by evacuated tube solar collectors, and connected to a water-based open cooling tower.
- A back-up vapour compression chiller (55 kW capacity), with air cooled condenser is installed. It provides chilled water to the cold distribution network. Without the absorption machine, this component would have required a cooling power of 90 kW.

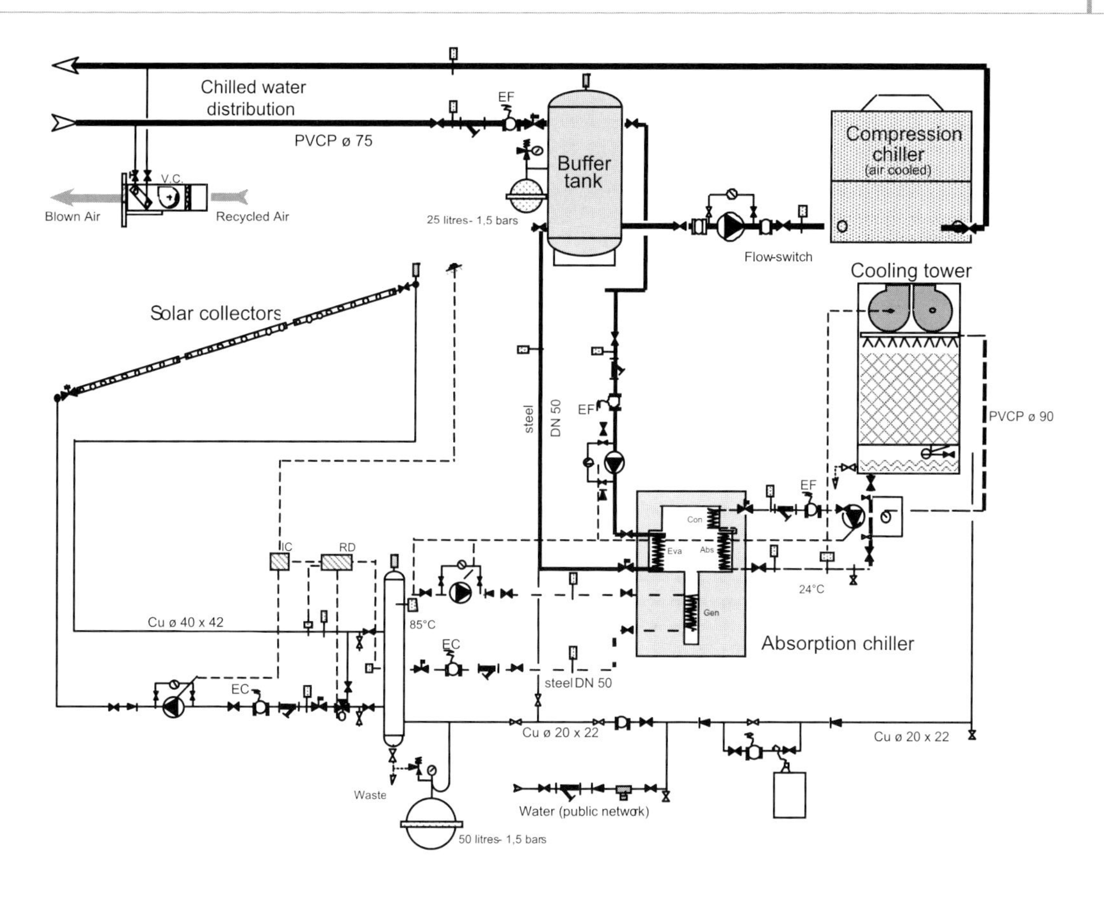

Figure 8.11
Plan of the solar assisted cooling system employing an absorption chiller.

This system was chosen for several reasons:

- The owner is responsible for environmental protection and was obliged to construct an exemplary building.
- The building should be an excellent demonstration project, employing reliable technology, it should operate trouble-free and offer comfortable working conditions.
- The cooling system and its controls should not be too sophisticated, in order to avoid the risk of frequent breakdowns. The reliability should be similar to traditional systems.
- The series configuration of the system, was chosen in order to ensure simplicity, reliability and safety.
- In Guadeloupe, the cooling demands exists throughout the whole year, as it does the solar energy.
- It is a real building, with real users, and not a laboratory. Thus, the experience gained is clearly more relevant to practical application.

8.2.3 Collector and solar system design

The heat supply for the absorption chiller is required in a temperature range of 85 - 95°C. This dictated the choice of evacuated tube collectors from the beginning of the project. Caraib'Froid Systems is the company who installed the plant. For the solar collector field the CORTEC evacuated tube collectors manufactured by J. Giordano Industries were selected. Preliminary system dimensioning was as follows:

- Cooling output: 30 kW
- Average COP: 0.70
- Input needed for the generator: 42.8 kW
- Reference solar energy supply: 700 W/m²
- Useful solar collector area needed: a total of 61.2 m² solar collectors.

The system is not equipped with full-size hot-water storage tank (i.e., there is no possibility to store heat for long periods). Instead, the collectors are connected to the generator of the absorption chiller by a small buffer tank (less than 100 litres). The cooling output (30 kW) was established so that it would always be slightly less than the load at all times during sunny periods. In this way, the absorption chiller can operate directly with the sun, and use all the energy supplied by the solar collectors for cooling. When the load is higher, the vapour compression chiller is used as back-up.

8.2.4 Energy performance

This system is expected to save a third of the electricity billed for air-conditioning every year, if compared to a conventional compression system, which would have a capacity of 90 kW:

- Cooling load: 68 681 kWh/year (68 kWh/m² per year)
- Solar cooling supply: 32 700 kWh/year
- Electricity saved: 10 900 kWh/year
- Primary energy saved: about 3 TOE/year
- Financial savings: 1 500 € /year
- Environmental impact: reduced emissions of 8 tons of CO_2 per year.

8.2.5 Economics

Figure 8.12 illustrates both the cost distribution and the initial cost breakdown for this solar-assisted cooling system. An additional cost of 18 000 € for system design should be added for the part concerning the solar cooling equipment. The solar collectors contribute by more than 40 % to the overall first cost.

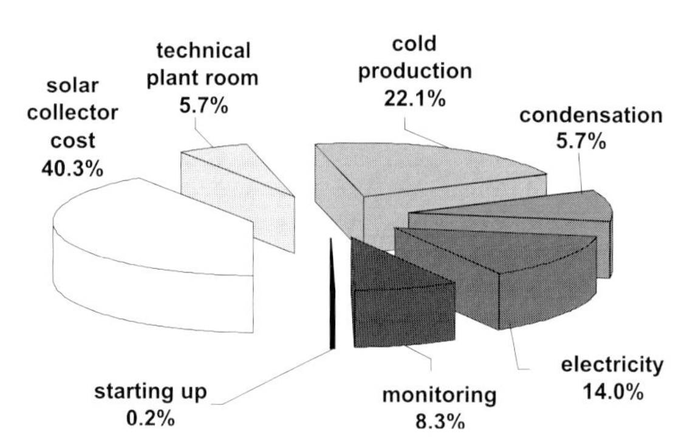

component	cost, €
solar collector cost	39 365.00
technical plant room	5 616.00
cold production	21 614.00
condensation	9 224.00
electricity	13 644.00
monitoring	8 093.00
starting up	203.00
total	97 759.00
specific cost, € per kW cooling power	2 794.00
Collector specific cost, € per m²	635.00

Figure 8.12
First cost distribution and the initial cost breakdown for the entire system.

8.3 The IHK desiccant cooling system in Freiburg, Germany

Figure 8.13
View of the IHK building in Freiburg/Germany.
A seminar room and a small cafeteria with large glazed areas to the east and the west are located on the top floor of the building.

GENERAL DESCRIPTION

The Chamber of Trade and Commerce for the south-west region of Germany (IHK Südlicher Oberrhein), is located in Freiburg in a building that was constructed in 1992 (see Figure 8.13). In the top floor of the building is a seminar room and a small cafeteria which is used when meetings or other events take place in the seminar room (see Figure 8.14). The seminar room and cafeteria were originally only equipped with an air ventilation system without air-conditioning function. However, conditions in the rooms during summer were very often uncomfortable, so it has been decided to install an air-conditioning system.

Following a detailed design process, a solar-assisted desiccant cooling system was installed, which is operated by a solar air collector without any back-up heat source.

Main information about the project:

Client:	IHK Südlicher Oberrhein/ Freiburg
Engineering:	Fraunhofer ISE
Solar collectors:	Grammer Solar + Bau GmbH
Desiccant Air-Handling Unit:	robatherm GmbH

8.3.1 Building and load

room	-	seminar room	cafeteria	total
floor area	m²	148	65	213
volume	m³	565	250	815
maximum number of persons	-	100	20	120
minimum required fresh air rate	m³/h	5000	1000	6000
minimum required air change rate	1/h	8.8	4.0	7.4

Table 8.2
Basic building and load data for the IHK seminar room and cafeteria.

Basic data for the building and the load are summarised in Table 8.2. The seminar room is designed for a maximum of about 100 occupants. Assuming a fresh air supply rate of 50 m^3/h per person, a fresh air flowrate of 5000 m^3/h results for the seminar room.

The cooling load is dominated by the conditioning of ventilation air, but internal loads due to occupants also play an important role. The windows are equipped with external shading devices to reduce solar gains during sunny days in summer. The cooling load at design conditions for the seminar room was assessed as 27.5 kW and for the cafeteria as 7.3 kW, i.e., the total cooling load is about 34.8 kW.

8.3.2 Air-conditioning concept and design of equipment

Figure 8.14
View of the interior of the small cafeteria with glazings directed to east and south direction.

Since high ventilation rates are required, it has been decided to install a desiccant cooling system to air-condition the seminar room and cafeteria; an all-air system could be used due to the high air flowrate needed to achieve the minimum required air change rates. The desiccant technique allows the use of low-temperature heat to drive the cooling process. This allows to use standard solar collectors, without requiring particularly high collector efficiencies.

In order to match the cooling load at the design point, i.e., completely occupied rooms under summer weather conditions, the design flowrate of the air-handling unit was chosen somewhat higher than the minimum required. The design figures of the air-handling unit are summarised in Table 8.3. The chosen desiccant cycle is a conventional ventilation cycle, where direct, indirect and combined evaporative cooling of the supply air stream can be implemented. The fans are equipped with frequency controllers allowing them to be able to adapt the ventilation rate according to the actual needs. Also, the humidifiers can be controlled in a range from 20 to 100 % efficiency. The cooling power, $\dot{Q}$, given in Table 8.3 is calculated as the enthalpy change of inlet air between the ambient air state and the supply air state, multiplied by the air mass flow

$$\dot{Q} = \dot{m}_{supply} (h_{amb} - h_{supply}) \qquad (8.1)$$

and the cooling power used to cover the room loads, $\dot{Q}_{int}$, is defined as the enthalpy change of air in the room, i.e., between supply and return air, multiplied by the air mass flow:

$$\dot{Q}_{int} = \dot{m}_{supply} (h_{return} - h_{supply}) \qquad (8.2)$$

Room	Unit	Seminar room	Cafeteria	Total
Maximum air-flow of air handling unit	m³/h	7800	2400	10200
Air change rate	1/h	13.8	9.6	12.5
Temperature, ambient air	°C	32.0		
Humidity ratio, ambient air	g/kg	12.0		
Temperature, supply air	°C	21.0		
Humidity ratio, supply air	g/kg	9.3		
Temperature, room air (= return air)	°C	26		
Humidity ratio, room air (= return air)	g/kg	10.5		
Cooling power (ambient ==> supply)	kW	40.5	12.5	53.0
Cooling power in the room (supply ==> return)	kW	16.8	5.2	21.9

Table 8.3
Design values for the desiccant air-handling unit.

8.3.3 Collector and solar system design

In a design study, it was investigated whether it is possible to install a solar air collector system without any back-up heat and thermal storage; a sketch of the designed system is shown in Figure 8.15 for operation in summer. In summer the ambient air is heated by means of the solar air collector and is used to regenerate the sorption wheel. Room return air is humidified and used in the indirect evaporative cooling process of the supply air stream. The system needs an additional fan compared to standard systems, in which return air is used for regeneration. During operation in winter the solar air collector is used for pre-heating of fresh air before it enters the heat recovery system. In case of high solar gains the heat recovery wheel is not operated. A schematic drawing of the system for operation during the heating season is shown in Figure 8.16. The rooms are equipped with a convective heating system. Therefore the ventilation air rate is limited to the actual minimum required in the heating season.

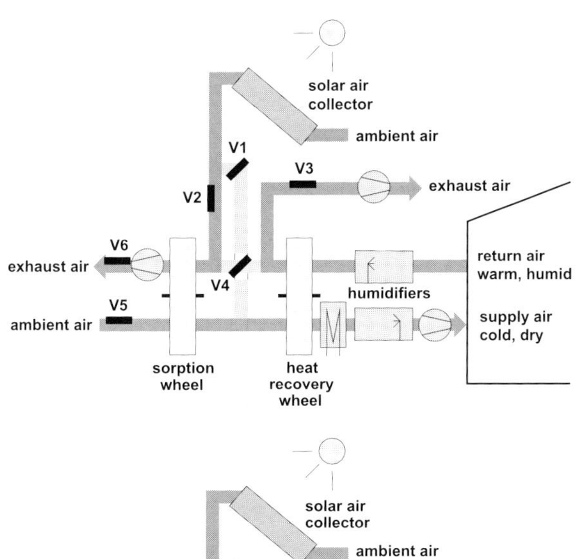

Figure 8.15
Schematic drawing of the desiccant cooling system with solar air collectors as the only heat source; settings for the operation during summer are shown (valves 1 and 4 closed; valves 2, 3, 5 and 6 open).

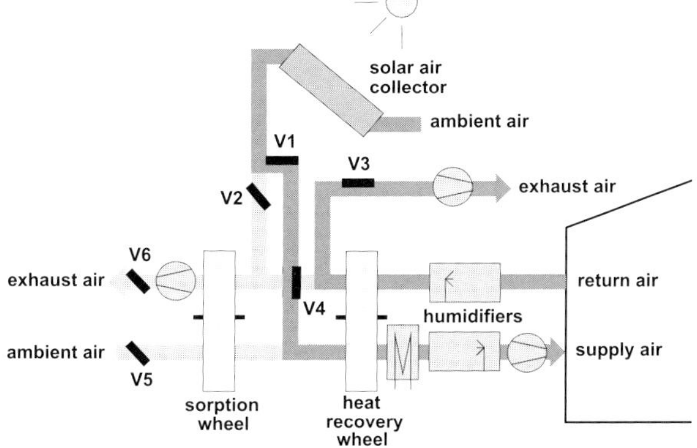

Figure 8.16
Schematic drawing of the desiccant cooling system with solar air collectors as the only heat source during operation in the heating season - solar radiation is highly available (valves 1 and 3 open; valves 2, 4, 5 and 6 closed).

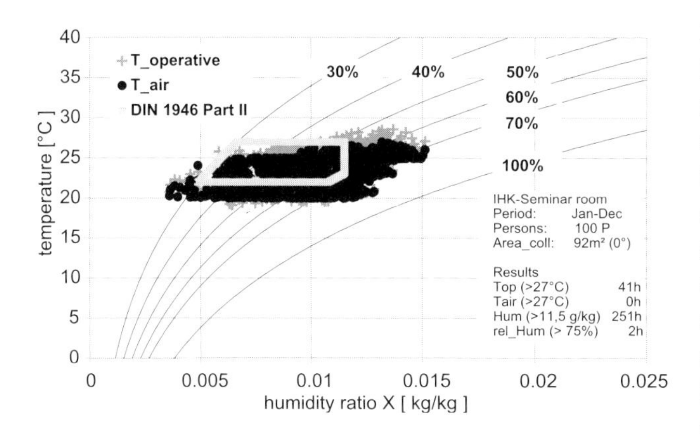

Figure 8.17
Result of system design. Each single point represents the indoor temperature versus indoor humidity ratio for one hour of the year. Grey crosses show the operative temperature and black dots the air temperature. The grey line indicates the comfort zone according to the German standard DIN 1946 part II.

In a system in which the solar collector is the only heat source it will most probably not be possible to match the cooling load at all times. Since no other heat or cold source is available, a design based on the solar contribution to the required cooling as a performance figure is not possible. The performance of the entire system has to be determined by analysis of the achieved indoor air states and assessing whether or not they are sufficient regarding indoor comfort. Therefore, a TRNSYS model was developed for the overall system consisting of solar collector, desiccant cooling air-handling unit and building. The annual performance of the overall system was calculated based on Test Reference Year data of Freiburg, Germany. For each hour of the year, the indoor air state was computed which results from the currently available solar heat and the resulting operation of the desiccant system. Results of simulations are shown in Figure 8.17.

Although the predicted room comfort exceeded the desired humidity and temperature levels for several hours of the year according to Figure 8.17, the client decided to follow this approach. It should be mentioned that the simulation was made for a worst-case scenario, assuming that the rooms are fully occupied during all working hours of the year. In reality for many hours the number of occupants is appreciably lower and some hours occur during the day when there is no use at all.

Figure 8.18
The solar air collector array on the roof on top of the seminar room.

The system was installed and commissioned in 2001 and has been operating satisfactorily since then. The air-handling unit and the solar collectors were manufactured by the German companies, robatherm and Grammer respectively. The solar air collector system installed on the roof on top of the seminar room is shown in Figure 8.18. Two arrays of 50 m² each are fixed directly onto the light-weight flat roof, without the need of a supporting structure. One array is directed to the east with a tilt angle of about 15° and the other one to the west with a tilt angle of 15°. The option of mounting the collectors directly onto the existing flat roof meant that installation was very simple.

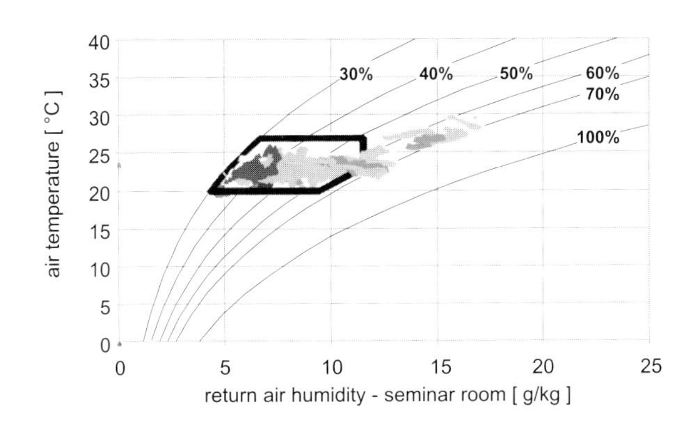

Figure 8.19

Performance of the system in the period July 2001 to January 2002; the figure can be directly compared with Figure 8.16 which shows the predicted values based on simulation.

8.3.4 Energy performance

The real performance of the system for the period between July 2001 and January 2002 is presented in Figure 8.19. It can be seen that the maximum room temperature corresponding to the comfort zone is exceeded only during a very few hours. The humidity value is exceeded more often, but except on two days the design value of 60 % relative humidity is met. The system performance on a typical summer day (example: August 15th, 2001) is shown in Figure 8.20.

According to the design study it was estimated that the system would save about 30 % of primary energy compared to a reference system. The absolute primary energy consumption of the reference system was calculated as 25 922 kWh and that of the desiccant cooling system with solar air collectors as the heat source as 18 162 kWh. An air-handling unit with heat recovery and a conventional compression chiller for dehumidification and temperature control was assumed, as reference system. The energy figures are valid for the entire year, including the heating season.

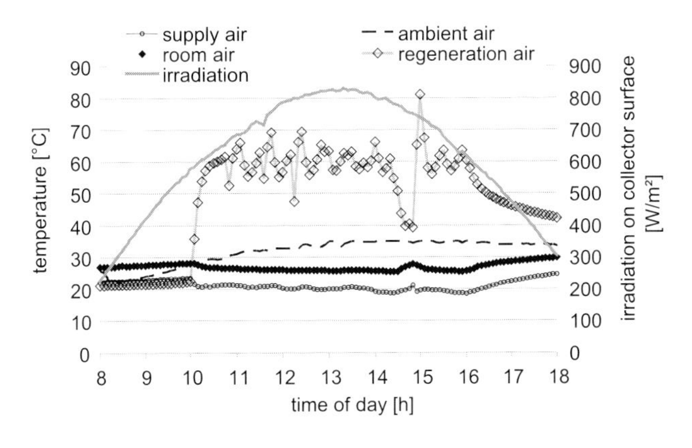

Figure 8.20

Temperatures on a typical summer day; the set value of 26°C for the room temperature only slightly exceeded.

8.3.5 Economics

Figure 8.21 shows both the distribution and the initial cost breakdown over the different cost categories. The total initial cost was € 210 000; all values are net cost without value-added tax (VAT). Calculation of specific cost figures leads to cost of 20.6 € per m³/h of nominal air flowrate for an air-handling unit with the four thermodynamic functions, i.e., heating, cooling, dehumidification and humidification. The specific cost in relation to the cooling power is 3961 € per kW of installed cooling power.

The dominant cost figure is the cost of the desiccant air-handling unit and its installation with about 69 %. The breakdown of this value is 46 % for the air-handling unit itself, 17 % for the installation and 6 % for the air ducts and their installation. Extra costs in the range of 5 % were due to the monitoring equipment, which required special sensors that are not necessary for control. The solar collector system contributes to the overall cost by only 10 %. This corresponds to a specific cost of 210 € per m² of collector gross area, including installation of the collector (228 € per m² of absorber area).

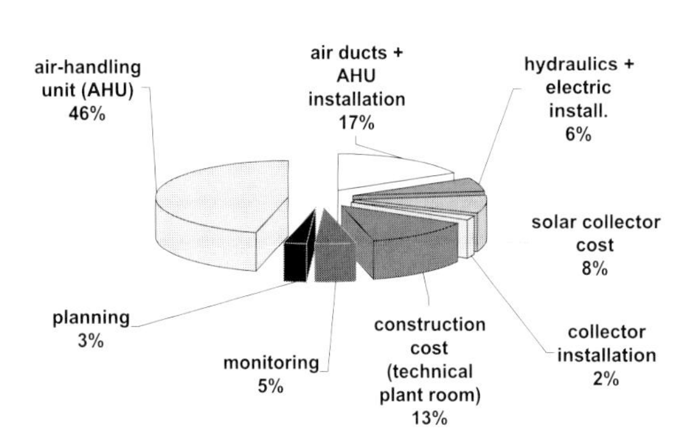

component	cost, €
air ducts + AHU installation	35 700
hydraulics + electric install.	12 600
solar collector cost	16 800
collector installation	4 200
construction cost (technical plant room)	27 300
monitoring	10 500
planning	6 300
air-handling unit (AHU)	96 600
total	**210 000**
specific cost, € per kW cooling power	3 962
specific cost, € per m^3/h nominal air-flow	20.59
specific cost, € per m^2 solar system	210.00

Figure 8.21
Cost distribution and cost breakdown for the different components and sub-systems of the IHK desiccant system; air-handling unit, total: 69 % and solar collectors, total: 10 %.

9 SUMMARY

Today, different types of technology are available to use solar thermal energy for air-conditioning of buildings. Several types of solar collectors, produced by numerous manufacturers in several countries, can be used to provide heat for the operation of thermally driven cooling equipment. Thermally driven sorption chillers, which use either a liquid or a solid sorbent to produce chilled water, are available on the market. However, the majority of them has a high cooling capacity. There are no available systems in the low cooling power range below 20 kW and only a few systems are commercially available in the cooling power range below 100 kW. Desiccant cooling air-handling units can be employed for conditioning the ventilation air using low-temperature heat from solar thermal collectors. Currently in use are mainly systems using solid sorbents fixed into the matrix of desiccant wheels. Systems using liquid sorption materials, which are well adapted for coupling with solar heat source, are also expected to be available on the market in the near future.

Basic Design Guidelines

For most thermally driven cooling equipment suitable for use in solar-assisted air-conditioning systems, the Coefficient of Performance (COP), that is the ratio between the produced cooling effect and the invested heat for this purpose, is noticeably below 1. This means that replacing a conventional air-conditioning system, which typically uses an electrically driven, vapour-compression chiller, by a solar-assisted system does not necessarily imply primary energy savings. Several design restrictions affecting solar-assisted air-conditioning systems result from this fact, as illustrated in Chapter 7 of this handbook:

- If a thermally driven cooling or air-conditioning system with a comparatively low COP is employed and a fossil-fuelled heat source is used as the back-up, a high solar fraction is necessary in order to achieve significant primary energy savings. An appropriate design of the solar system, i.e., suitable dimensioning of the solar collector and system-integrated energy storage, is necessary for this purpose.
- Systems employing thermally driven cooling equipment with a high COP lead to energy savings even at comparatively low solar contributions to the required heat for driving the system.
- In cases where a back-up system is needed either a second heat source, such as a back-up burner, to drive the thermally driven cooling equipment or a conventional chiller may be employed. The latter option may be appropriate if a large overall amount of cooling power is needed. In this case, the solar system mainly serves to reduce electric energy consumption as well as peak electricity loads.
- Solar-thermally autonomous systems that do not use any conventional heat source or back-up on the cold side, may also be used. In these cases, energy savings are always achieved but there is no guarantee of meeting the cooling loads and maintaining the indoor climate within the comfort conditions.
- In any case, the use of the solar collector field should be maximised through the exploitation of the solar heat source to match other loads such as space heating or domestic hot water production. Particularly in climates with high cooling loads during summer, the solar system can also contribute significantly to meet the heating loads during winter. For the examples of loads described in Chapter 6, the solar fraction for heating is in the same range as the one for the air-conditioning in summer at all investigated sites in the Mediterranean area. In these cases, the solar thermal system contributes considerably to achieve indoor thermal comfort throughout the year.

R&D Activities

Solar-assisted air-conditioning technology is still in an early stage of development. However, increased efforts in technological development will help to increase the competitiveness of solar-assisted air-conditioning approaches in the future. The following major trends are observed:

- Many research and development activities are ongoing, with the objective of providing thermally driven cooling equipment in the low cooling power range, for example, less than 20 kW. New equipment in this power range will open up new market opportunities for solar-assisted air-conditioning.
- Work on advanced absorption machines is advancing in two main directions. On the one hand projects aim on the development of single-effect cycles with increased COP-values at low driving temperatures, thus increasing the overall efficiency of solar-assisted cooling with standard solar collectors. On the other hand activities focus on the development of double-effect or even triple-effect cycles, which lead to a significant increase of the COP-values. However, there is no field experience on operating these machines with solar thermal collectors, since much higher driving temperatures are needed compared to single-effect cycles. Other types of solar collectors, such as concentrating collectors - either with or without sun tracking, depending on the concentration factor - are necessary to achieve sufficiently high collector efficiency at these high temperature levels.
- The main limitations of adsorption chillers today are due to the limited heat transfer between the granular sorption material and the water that flows through the heat exchanger in the compartments containing the sorption material. For this kind of systems, several new approaches, such as heat exchangers coated with the sorption material, are studied in research projects and results look promising. Distinct improvements in the power density and the thermal performance of adsorption chillers can be expected as a result of this work.
- Steam jet cycles are another promising candidate for closed cycles to produce chilled water. Research work focuses on the development of new schemes and new working fluids consisting of fluid mixtures, which allow low-temperature heat from solar thermal collectors to be used to drive the systems. Particularly under part load conditions, promising performance characteristics are expected.
- Open cooling cycles strongly profit from cooling of the sorbent during the sorption process. Research activities on cooled open sorption cycles are ongoing for both types of materials, i.e., solid and liquid sorbents. This enables higher dehumidification rates at the same or even lower driving temperatures, when compared to the desiccant processes used today. Processes employing liquid materials provide another advantage, namely to use the concentrated solution as a storage medium for air-conditioning purposes with a high energy density. This allows a favourable decoupling of the sorbent regeneration using solar energy and the air dehumidification.

All these and other future developments will increase the potential for using solar thermal energy for air-conditioning. However, it is not necessary to wait for the future, since - as documented in this handbook - already today there are many possible solutions and available equipment.

In addition, more knowledge at the systems level is also needed in order to increase efficiency and reliability. Hopefully, pioneers among planners, engineers and end-users will continue to use solar thermal energy for air-conditioning, even if greater effort is necessary today during planning and often also during operation. Their experience gained in future installations will contribute to a continued learning process and help to make this technology increasingly successful. Cost reductions due to solutions with a higher level of standardisation and shortening of the design process will also be achieved. System operation will become more reliable and the effort for maintenance will be reduced by continued equipment development. Overall, it is a highly exciting and challenging goal to use the energy from the sun to provide cooling and air-conditioning which is mainly provoked by the same source - the sun.

REFERENCES

/1.1/ Commission of European Communities (2001), Green Paper - Towards a European strategy for the security of energy supply. Brussels/Belgium.

/1.2/ Adnot J. - co-ordinator (1999). Energy Efficiency of Room Air-Conditioners (EERAC),. Study for the Directorate-General for Energy (DGXVII) of the Commission of the EC, Final Report.

/1.3/ Mouchot, A.: 'La chaleur Solaire et ses Applications Industrielles'. German translation 'Die Sonnenwärme und ihre industriellen Anwendungen', Olynthus Verlag, Oberbözberg, 1987.

/1.4/ Löf G. - editor, (1993). Active Solar Systems, Volume 6 in Solar Heat Technologies series. (Bankston C.A. - editor) MIT Press, Cambridge, MA/USA.

/1.5/ Sayigh A.A.M. and McVeigh J.C. – editors, (1992). Solar Air Conditioning and Refrigeration. Pergamon Press, Oxford/UK.

/1.6/ Santamouris M. and Asimakopoulos D., - editors (1996). Passive Cooling of Buildings, James & James, London/UK.

/2.1/ Lecture book on Indoor Climate Control (1996), TU Delft/NL.

/2.2/ ASHRAE (2000). Handbook 2000 HVAC systems and equipment - SI edition. American Society of Heating, Refrigeration and Air-Conditioning, Atlanta GA/USA.

/2.3/ Eurovent Standard 6/3 (1996). Thermal Test method for Fan-Coil Units. Eurovent/Cecomaf, Paris/France.

/2.4/ European Standard: Floor heating - Systems and components (/EN-1264-1,2,3 1997 and EN-1264-4 2001/).

/3.1/ Franzke U. (2001). Personal communication, Institut für Luft- und Kältetechnik (ILK) Dresden/Germany.

/3.2/ Leaflet Yazaki Japan, date Dec. 27, 2001.

/3.3/ Marktübersicht Absorptionskältenanlagen - Angebot und Anbieter. Arbeitsgemeinschaft für sparsame und umweltfreundlichen Energieverbrauch (www.asue.de)

/3.4/ ASHRAE (2002) Absorption Cooling, Heating and Refrigeration Equipment, in Handbook 2002 Refrigeration, Chapter 41. - SI edition. American Society of Heating, Refrigeration, and Air-Conditioning, Atlanta GA/USA.

/3.5/ Performance data of adsorption chillers of type MYCOM (manufacturer: Mayekawa Mfg. Co., Ltd.). Received from Albring Industrievertretung GmbH, Alsbach-Haehnlein, Germany, 2002.

/3.6/ Wurm J., Kosar D. and Clemens T. (2002). Solid Desiccant Technology Review, Bulletin of the IIR, Vol. 3, p. 2-31.

/4.1/ Duffie J. A. and Beckman W. A. (1991). Solar Engineering of Thermal Processes – 2nd edition. John Wiley & Sons, Ltd. New York.

/4.2/ Rommel M., Gombert A., Koschikowski J., Schäfer A. and Schmitt Y. (2003). Which Improvements can be achieved using single and double AR-glass covers in flat-plate collectors?, Proceedings of European Solar Thermal Energy Conference, ESTEC 2003, 26-27 June 2003, Freiburg, Germany.

/4.3/ Weiss W. and G. Faninger (2002). Solar thermal collector market in IEA member countries. IEA Solar Heating and Cooling Programme.

/4.4/ LTS- Collector Catalogue 2002, Institue für Solartechnik SPF Rapperswil. BFE Bundesamt für Energie, Bern/Switzerland.

/4.5/ Farinha Mendes J. (2003), Personal communication, dept. LECS at INETI - Instituto Nacional de Engenharia e Tecnologia Industrial, Lisbon/Portugal (efficiency values).

/4.6/ Personal communication (2003) Ao Sol-Energias Renováveis, Lda, Samora Correia/Portugal (cost figures).

/4.7/ Lane G. A. (1986). Phase Change Materials, in Solar Heat Storage: Latent heat Material, Lane G.A. – editor Volume II, CRC-Press, Boca Raton, Florida/USA.

/5.1/ Quinette J.Y. (2002) - Personal communication - TECSOL S.A. – Perpignan/France.

/5.2/ Collier R.K.J. (1997). Desiccant dehumidification and cooling systems assessment and analysis. Pacific Northwest National Laboratory, Richland, Washington/USA.

/5.3/ Pesaran, A. A., Penney, T. R. and Czanderna, A. W. (1992). Desiccant cooling: State-of-the-art assessment, NREL, Golden CO/USA.

/5.4/ Hindenburg C. and Henning H.-M. (2002) - Systemlösungen und Regelungskonzepte von Solarunterstützten Klimatisierungssystemen, Teil 2 - Sorptionsgestützte Klimatisierung - HLH Heizung Lüftung/Klima Haustechnik Bd. 53- pages. 83-90 Number 6.

/5.5/ Commission of European Communities (2000) - DIRECTORATE-GENERAL FOR ENERGY - Energy in Europe, 1999 Annual energy review - Special Issue 2000.

/6.1/ METEONORM -Global meteorological database for solar energy and applied climatology - PART I, II, III: REVIEW AND SOFTWARE - Meteotest - 1999 - V. 4.00.

/6.2/ POLYSUN - simulation software for solar thermal systems of small and medium size. Developed at: Institut für Solartechnik SPF, Hochschule für Technik, Rapperswil/Switzerland. Actual version 3.3.5, 2002; http://www.solarenergy.ch .

/6.3/ http://www.eere.energy.gov/buildings/tools_directory/

/6.4/ Henning H.M., Balaras C.A.; Grossman G., Infante-Ferreira C.A., Podesser E., Wang L. and Wiemken E. (2003). Solar assisted air conditioning - a new market for solar thermal energy in Europe. Proceedings of European Solar Thermal Energy Conference, ESTEC 2003, 26-27 June 2003, Freiburg, Germany.

/6.5/ Henning, H.M. and Wiemken E. (2003). Solar-Assisted Air-Conditioning of Buildings - An Overview. Proceedings of ISES Solar World Congress 2003- Solar Energy for a Sustainable Future, Gothenborg/Sweden, June 2003.

/6.6/ TRNSYS - A Transient System Simulation Program - a program developed at the SEL - Solar Energy Laboratory, University of Wisconsin, Madison/USA. Actual version 15.3, 2003 - http://sel.me.wisc.edu/TRNSYS/Default.htm.

/6.7/ http://de.expertcontrol.com/carnot.htm

/6.8/ http://www.colsim.org/

/6.9/ http://www.smilenet.de

A1 APPENDIX 1

Description of the reference buildings

Three typical reference buildings have been selected to define the corresponding loads, namely:

- Hotel
- Office
- Lecture room

In thermal building simulations, a number of pre-defined input parameters and variables (building layout, occupancy, internal gains, ventilation, weather data, etc.) have to be specified. The interpretation and thus the accuracy of the simulation results depends on the validity of these calculation assumptions.

Some essential calculation assumptions were already presented in Table 6.1 and discussed in Chapter 6. Additionally the pre-defined U-values for some construction building components are used for all three building types. Table A1.1 contains representative U-values for different wall types and windows that are characteristic for the sites given the respective meteorological conditions.

The following section presents details of all the relevant data and assumptions made for simulation of the various reference buildings.

U-values		Freiburg Copenhagen	Perpignan	Madrid	Palermo	Athens	Merida
external wall	[W/m²K]	0.350	0.320	0.591	1.260	0.600	0.704
roof	[W/m²K]	0.170	0.330	0.318	0.480	0.500	0.358
ground floor	[W/m²K]	0.350	0.422	0.356	0.845	1.500	1.500
floor	[W/m²K]	0.360	0.230	0.743	0.590	1.500	1.500
internal wall	[W/m²K]	0.370	0.292	1.572	1.250	1.500	1.500
window	[W/m²K]	1.100	1.400	3.900	4.130	3.700	5.319
frame	[W/m²K]	2.000	2.270	2.260	2.000	2.260	3.000
frame share of window	[m²/m²]	0.2	0.2	0.2	0.15	0.2	0.2
g-value of window	[-]	0.598	0.589	0.768	0.8	0.8	0.8

Table A1.1
U-values for different wall types and windows, characteristic for the chosen meteorological sites.

A1.1 Definition of the reference hotel building

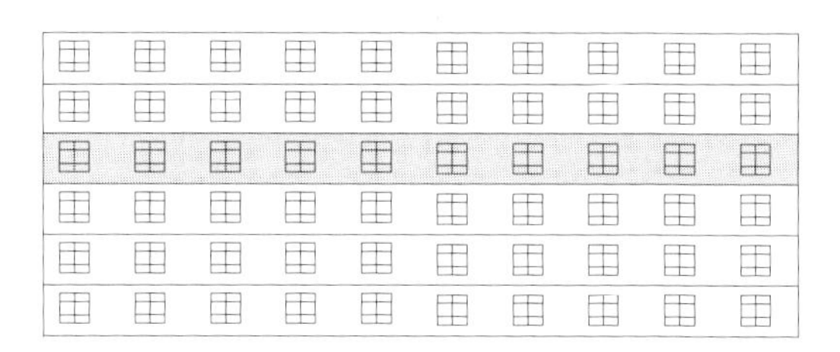

Figure A1.1
Facade of the hotel building; the shaded area identifies the floor for which the simulation was performed.

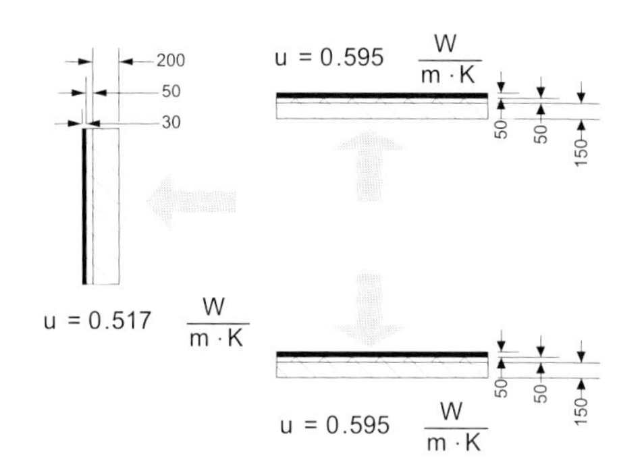

Figure A1.2
Construction of external building elements; all dimensions are in millimetres; the construction materials are identified in Table A.1.2 - the values refer to the calculation for Freiburg, Germany.

Architectural design

The reference object is a free-standing, six-storey hotel building (Figure A1.1). The building is oriented along the east-west axis with an internal access corridor in the centre. The effective floor area on one storey, including the areas covered by the internal walls and the access areas, is 642.6 m². The glazed area on the north and south facade amounts to 25 % of the facade area and 4% on the east and west facade. The simulations were performed for only one storey of the building.

Construction

The construction of the building envelope (external walls, roof and floor slab between ground floor and basement) is reduced to the base slap (storage layer) and the insulation layer.

The various materials, layer thicknesses and energy performance data describing the hotel simulated for Freiburg, Germany are listed in Table A1.2. The calculation were performed taking into account representative standards values according to the site and the respective meteorological data - see Table A.1.1.

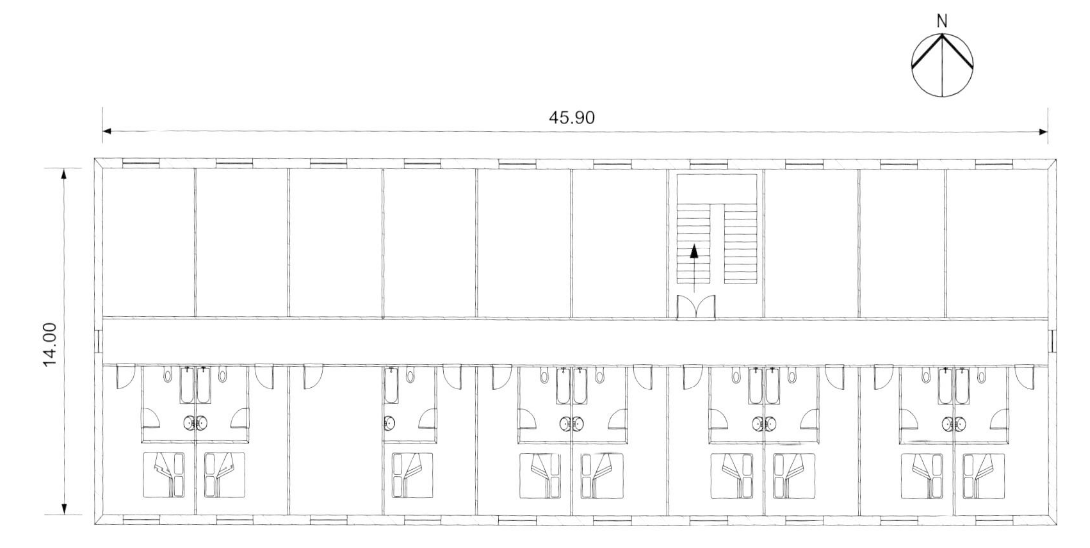

Figure A1.3
Geometry and dimensions of reference hotel building - all dimensions are in metres.

Geometry and dimensions

Figure A1.3 shows the geometry and dimensions of the considered floor of the hotel building.

Internal Walls

Internal walls are not considered.

Windows

One fourth of the total surface area on the north and south facade is glazed. On the west and on the east facade, the corresponding value is only 4%of he total surface. Table A1.1 contains the assumed values for thermal and transmission properties of the windows and frames.

External Shading

The effect of the blinds can be modelled to depend on the incident radiation of the sun. In general, the external blind shades the window is down, if direct radiation falls.

Internal gains

Occupancy

In the case of full occupancy 19 people are present simultaneously in the floor of the building. At noon, there are only two hotel employees on each floor. The same number of occupants is in the building on Saturdays and Sundays. Numerical values for the overall occupancy profile are displayed in Figure A1.4.

Assembly	Layer	Thickness [mm]	Density [kg/m³]	Thermal conductivity [W/m K]	U-value [W/m² K]
External wall	Plaster	30	200	0.06	2
	Insulation	50	180	0.045	0.9
	Concrete	200	1 800	1.3	6.5
	Total				**0.517**
Floor slab	Covering	50	2 000	1.4	28
	Insulation	50	30	0.04	0.8
	Concrete	150	2 000	1.6	10.66
	Total				**0.595**
Window					1.1

Table A1.2
Construction components of reference hotel building calculated for Freiburg, Germany.

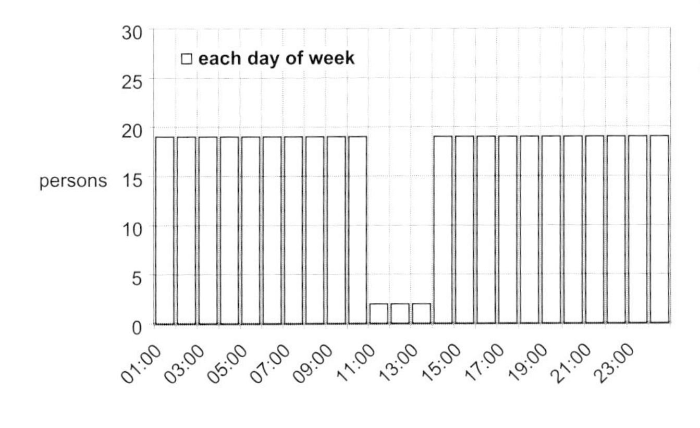

Figure A1.4
Occupancy profile for each day of the week.

As soon as a person is present in the building it is assumed that this person will dissipate heat to the surroundings at a rate of 115 W for the hotel guests and 190 W for the employees. This value is in accordance with the german VDI 2078 standard and is based on the following scenario:

Degree of activity - guests
sensible heat 70 W/person
latent heat 45 W/person

Degree of activity - employees
sensible heat 85 W/person
latent heat 105 W/person

An air change rate of 0.5 volumes per hour due to leakages in the building envelope is assumed. The resulting net value for the hourly volume air change in the entire building is 900 m³ per hour per storey.

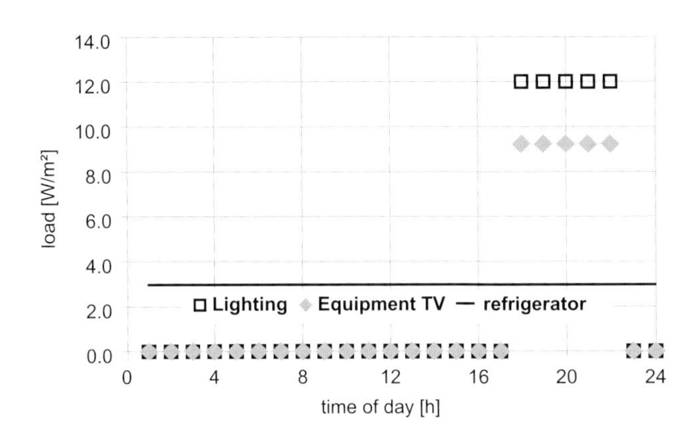

Figure A1.5
Daily profile of the internal heat gains.

Two main sources of equipment-related heat dissipation are taken into account: hotel equipment (i.e., TV, refrigerators) and artificial lights. The daily profile is shown in Figure A1.5.

The rate of heat dissipation from hotel equipment depends on its operational status and is thus correlated to the occupancy profile in the rooms. It is assumed that all rooms are equipped with a refrigerator with a heat dissipation rate of 100 W and a TV with a heat dissipation rate of 250 W per unit.

The lighting in all cases is correlated to the usage periods of the rooms. In the model, the term 'usage periods' applies to the time interval between 18 and 22 hours. The lighting, which is assumed to be generated by standard lamps (heat dissipation: 12 W/m²), is switched on during this time

A1.2. Definition of the reference office building

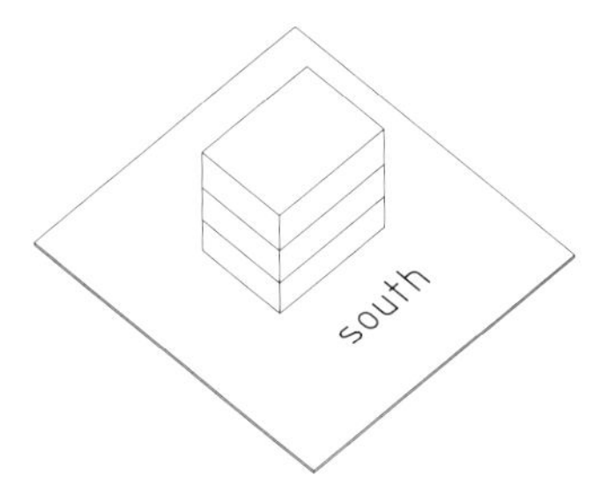

Figure A1.6
Orientation of reference office building.

Architectural design

The chosen reference object is a three storey office building with a basement (see Figure A1.6). The building is oriented along the east-west axis and it is designed to be internally accessed. The floor area on one level including the areas covered by internal walls and access facilities amounts to 309.9 m². The glazed area on the east facade amounts to 10 % of the wall surface area. On the south and on the north facade, the corresponding value is 37.39 % whereas on the west facade, the building is assumed to be without windows. Based on the selected physical dimensions of the object, the characteristic length (= volume of building/area of building envelope) amounts to 2.63 m.

Construction

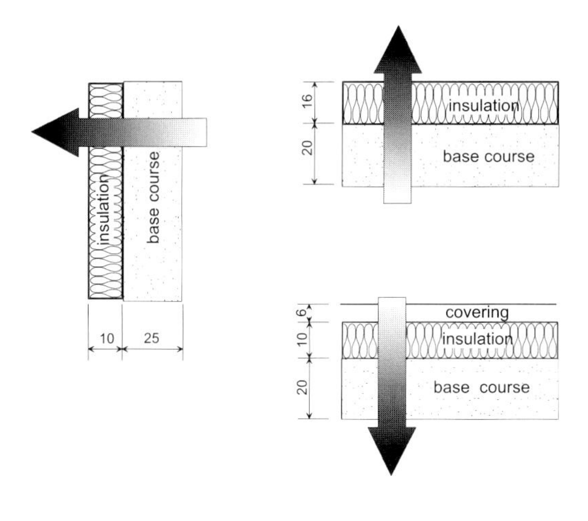

Figure A1.7
Constructional composition of building envelope.

The construction of the building envelope (external walls, roof and floor between ground storey and basement) is reduced to the base slab (storage layer) and the insulation layer. The code of practice with regard to structure analysis and acoustics determines the thickness of the two layers. Other values specifying the layers (density, specific heat capacity, thermal conductivity) follow the characteristics of commonly used materials. The resulting U-values are in accordance with Austrian standards. For the upper surface of the floors, a covering layer is assumed to supplement the base slab and the insulation layer. Numerical values are summarized in Table A1.3.

GEOMETRY / DIMENSIONS

Figure A1.8 displays the geometry and dimensions of the office building. The resulting areas and volumes are summarized in Table A1.4.

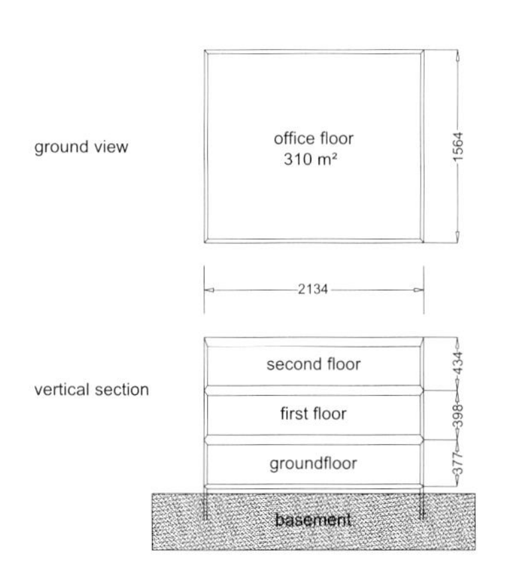

Figure A1.8
Geometry and dimensions of the reference office building.

INTERNAL WALLS

The dimensions of the internal walls of the reference office are identical on each floor. The internal walls are 80 m long and 3.18 m high, leading to a total internal wall area of 254.4 m² per floor and side.

OCCUPANCY

The net floor area on each level of the building has been partitioned into different types of rooms (Table A1.5). In those areas which represent offices or rooms for customer reception, one person is assumed to occupy 15 m² floor area. In the case of full occupancy this assumption leads to a total of 32 people being simultaneously present in the building. On Saturdays, working activities and a correspondingly reduced occupancy profile are restricted to the first floor whereas the other floors are assumed to be unoccupied. On Sundays, no people are assumed to be inside the building. Numerical values for the overall occupancy profile are displayed in Figure A1.9.

assembly	layer	thickness	density	thermal conductivity	specific heat capacity	U - value of component
		[m]	[kg/m³]	[W/mK]	[kJ/kgK]	[W/m²K]
	Insulation	0.100	17	0.038	0.830	0.35
	base	0.220	2 400	2.300	1.080	
	Σ	0.320				
roof	Insulation	0.160	30	0.032	0.840	0.17
	base	0.200	2 400	2.300	1.080	
	air	0.380	-	-	1.007	
	bottom	0.020	100	0.035	0.840	
	Σ	0.760				
floor slab (ground floor-basement)	top	0.080	2 400	2.300	1.080	0.35
	insulation	0.100	30	0.038	0.830	
	base	0.200	2 400	2.300	1.080	
	Σ	0.380				
floor slab	top	0.040	700	0.130	1.000	0.36
	air	0.090	-	-	1.007	
	insulation	0.050	30	0.038	0.830	
	base	0.200	2 400	2.300	1.080	
	air	0.400	-	-	1.007	
	bottom	0.020	100	0.035	0.840	
	Σ	0.800				
internal wall	base	0.015	900	2.100	1.050	-
	insulation	0.100	10	0.039	0.840	
	base	0.015	900	2.100	1.050	
	Σ	0.130				

Table A1.3

Constructional components of reference office building assumed for Freiburg, Germany.

	net values						gross values						
	dimensions			floor area (excl. area occupied by walls)	facade area (north and south)	volume	dimensions			floor area	facade area (north and south)	facade area (east and west)	volume
	L [m]	W [m]	H [m]	L x W [m²]	L x W [m²]	L x W x H [m³]	L [m]	W [m]	H [m]	L x W [m²]	L x H [m²]	W x H [m²]	L x W x H [m³]
ground floor	20.66	15.00	3.18	309.90	65.70	985.48	21.30	15.64	3.98	333.13	84.77	62.25	1 325.87
first floor	20.66	15.00	3.18	309.90	65.70	985.48	21.30	15.64	3.98	333.13	84.77	62.25	1 325.87
second floor	20.66	15.00	3.18	309.90	65.70	985.48	21.30	15.64	4.34	333.13	92.44	67.88	1 445.78
total				**929.70**	**197.10**	**2 956.44**				**999.39**	**261.98**	**192.38**	**4 097.52**

Table A1.4

Areas and volumes of the reference office building.

Office Equipment

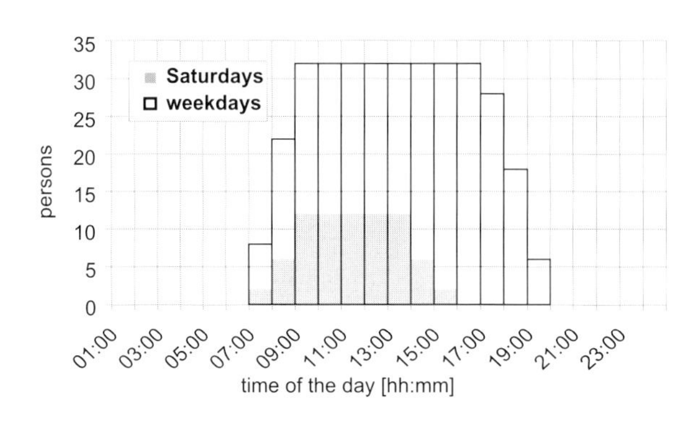

Figure A1.9

Occupancy profile for different days of the week.

The rate of heat dissipation from office equipment depends on its operational status and is thus correlated to the occupancy profile in the offices. It is assumed that 70 % of the work places are equipped with a PC (including a monitor) with a heat dissipation rate of 230 W per unit.

Artifical Lighting

The floor area is divided into two sections. One section is assumed to be illuminated only by daylight and the other one only by artificial lighting. The lighting in the latter case is correlated to the working hours and thus to the occupancy profile. In the model, the term 'working hour' applies to all time intervals in which at least one person is present in the office rooms. It is assumed that the lighting is on during these times and that energy saving lamps are used (heat dissipation: 2 W/m²).

WINDOWS

room type	floor area [m²]
offices	330
meeting rooms	60
access areas (corridors, staircases, etc.)	180
social rooms	40
lavatory	45
room for customer reception	150

Table A1.5
Floor areas of different sections of the reference office building.

Ten percent of the total surface area of the east facade is glazed. On the south and on the north facade, the corresponding value is 37.4 %, whereas the west facade is assumed to be without windows - see Table A1.6. The quality of the glazing and the window frame is determined for thermal and transmission properties in accordance to specific standards used in practice at the different meteorological sites - see Table A1.1.

Specific fractions of the incident beam radiation are assigned to various surface areas in a zone. The respective values are given by:

25%	floor
50%	ceiling
25%	internal walls

	facade area (north and south)	glazed area	glazed area incl. 27.5% frame area	facade area (north and south)	glazed area	glazed area incl. 27.5% frame area
	[m²]	[m²]	[m³]	[m²]	[m²]	[m²]
ground floor	84.77	32.15	40.99	61.93	6.30	8.04
first floor	84.77	32.15	40.99	62.25	6.30	8.04
second floor	92.44	32.15	40.99	67.88	6.30	8.04
total	**261.98**	**96.45**	**122.97**	**192.06**	**18.90**	**24.12**

Table A1.6
Facade and windows areas of the reference office building.

A1.3 Definition of the reference lecture room

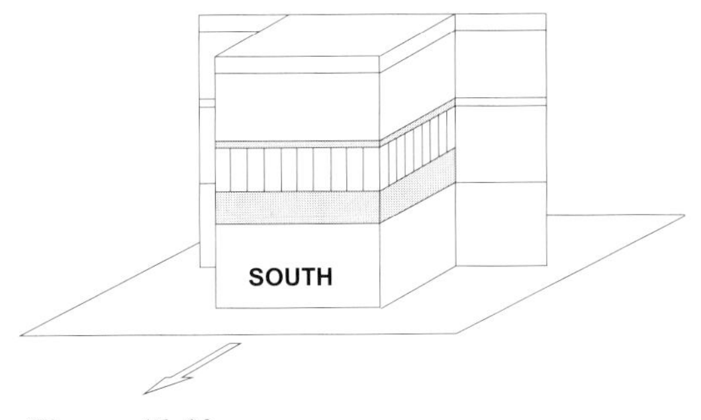

Figure A1.10
Axionometric diagram of reference office building; the shaded area identifies the room for which the simulations were made.

Architectural design

The reference lecture room is designed as a single-zone room. From the construction point of view, the lecture room is integrated in a three-storey building - so no thermal effects due to a basement or roof construction were taken into account. Figure A1.10 shows the geometry and relative dimensions of this building type. This so-called 'sandwich room' has a simple rectangular geometry - the floor area is 12 m width by 18 m length - leading to a total floor area of 216 m². Three of four walls are defined as external walls; their facade are orientated to the south, west and east. The glazed area of all facade amounts to 50 % of the walls' surface area. This means that a total glazed area of 63 m² is considered in the load simulation. An external shading device with a shading efficiency of 70% is operated separately if the beam radiation incident on the corresponding facade exceeds 200 W/m². With the exception of the external walls mentioned above, the other walls act thermally as an adiabatic boundary, i.e. no thermal conduction effects are active.

construction	layer	thickness	density	thermal conductivity	specific heat capacity
		[m]	[kg/m³]	[W/m K]	[W/m² K]
external Wall:	concrete	0.100	2 100	7.326	0.920
	mineral wool	0.060	75	0.169	0.840
	plaster	0.025	1 300	1.620	1.050
internal Wall	plaster board	0.009	900	0.760	1.000
	wooden board	0.012	1 000	0.610	1.000
	insulation	0.100	80	0.160	0.900
	wooden board	0.012	1 000	0.610	1.000
	plaster board	0.009	900	0.760	1.000
ceiling:	tin	0.001	7 800	208.800	0.480
	mineral wool	0.020	75	0.169	0.840
	concrete	0.120	2 100	7.326	0.920
	mineral wool	0.020	75	0.169	0.840
	floor cement	0.030	2 200	5.040	1.050
floor	floor cement	0.030	2 200	5.040	1.050
	mineral wool	0.020	75	0.169	0.840
	concrete	0.120	2 100	7.326	0.920
	mineral wool	0.020	75	0.169	0.840
	tin	0.001	7 800	208.800	0.480

Table A1.7
Construction components of the reference lecture room assumed for Freiburg, Germany.

Construction

The lecture room envelope is composed of the following wall types: external walls, internal wall, ceiling and floor. The various materials, layer thicknesses and energy performance data for Freiburg, Germany are listed in Table A1.7. According to Table A1.1, the U-values of each considered wall are adapted to the specific standards used in practice at the different meteorological sites.

Occupancy

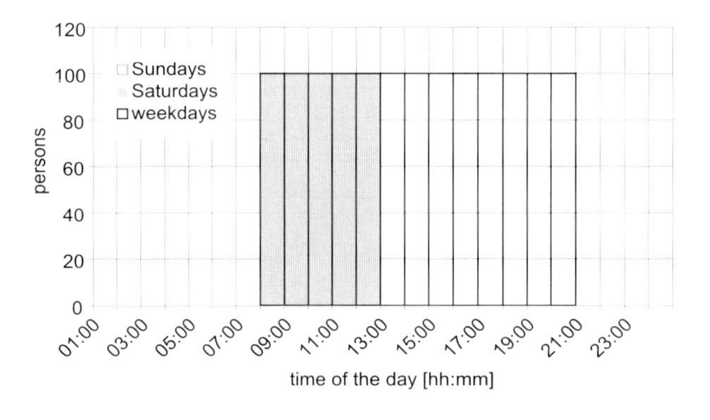

Figure A1.11
Occupancy profile for weekdays, Saturdays and Sundays for the reference lecture room.

Internal Loads

Lecturing activities were assumed according to a weekly occupancy profile without any variations according to seasons or vacation periods. An occupancy of 100 persons is assumed continuously during weekdays from 8:00 a.m. to 8:00 p.m. without taking any break into account. On Saturdays, the same number of persons are present from 8:00 a.m. up to 12:00 noon.

No people are assumed to be inside the building on Sundays. The corresponding weekly human occupancy profile is given by Figure A1.11. As soon as a person is present in the building, it is assumed that this person will dissipate heat to the surroundings at a rate of 120 W. This value is in accordance with the German VDI 2078 and DIN 1946-II standards and is based on the following scenario:

Degree of activity - seated at rest
sensible heat: 85 W/Person
latent heat: 35 W/Person

Infiltration

An air change rate between 0.2 and 0.5 volumes per hour is assumed due to leakages in the building envelope. Different infiltration rates are defined, depending on the considered climates. Madrid is calculated with an infiltration rate of 0.5, all the other sites are characterised by 0.2. The resulting net value for the hourly volume of air change due to infiltration in the entire lecture room is 130 m^3 per hour for all sites with the exception of Madrid (324 m^3 per hour).

TECHNICAL EQUIPMENT

Regarding the heat dissipation by technical equipment for the lecture room, only the artificial lighting is taken into account. There is no lighting control implemented to use daylight or artificial lighting following a certain control criteria. The lighting, which is assumed to be supplied by incandescent bulbs (40% heat transfer by convection) is switched on while the room is occupied. A specific heat dissipation of about 5 W/m^2 is calculated for the lighting.

WINDOWS

Fifty percent of the total surface area on the south, east and west facade is glazed. The quality of the glazing and the window frame is adapted to the regional standards for each site, similarly to the U-values of the walls . Table A1.1 contains the assumed values for thermal and transmission properties of the windows and frames.

EXTERNAL SHADING

Overheating due to the solar gains through the glazed areas is reduced by external shading devices. Control of the external shading is modelled as a function of the incident solar radiation falling on the respective facade (e.g., beam radiation higher than 200 W/m^2, external shading is active and only 30 % of the solar radiation is passing the window).

A2 APPENDIX 2

Parameters used for the examples of performance analysis (Chapter 7.2.1)

The following parameters were used in Chapter 7.2.1 for the example of performance analysis (office building in Madrid).

Components of the heat supply system			
Solar Collector			
Type	stationary CPC collector	flat-plate collector	
c_0 (optical efficiency), -	0.94	0.799	
c_1 (linear heat loss coefficient), W/(m²K)	2.2	3.12	
c_2 (quadratic heat loss coefficient), W/(m²K)	0.033	0.0199	
specific collector cost (including support structure), €/m²	400	300	
electric consumption of pumps in solar circuit, W	250		
temperature loss in heat exchanger (primary/secondary circuit), K	5		
Hot water buffer tank			
heat loss coefficient, W/(m²K)	0.8		
maximum storage temperature, °C	95		
specific costs of storage, €/m³	500		
Back-up Heater (Gas burner)			
efficiency (heating power/fuel), -	0.95		
specific costs back-up heating, €/kW	400		
operation temperature of the heating system (winter), °C	45		
electricity consumption of pumps in heating circuit, W	250		
Components of the cooling system			
Thermally driven chiller			
Type	absorption	adsorption	compression
effective average COP	0.65	0.6	3
operation temperature of the chiller	90	75	-
specific costs of the chiller, €/kW	400	850	300
Cooling tower			
electric power consumption per kWh of cooling, kWh $_{el}$/kWh$_{cooling}$	0.03		-
water consumption (factor in relation to evaporated water)	1.5		-
specific cost of cooling tower, €/kW $_{cooling}$	35		-
electricity consumption of pumps in cooling circuit, W	500		200
Others parameters			
initial cost			
overall installation cost, €	20 000		
cost of control system, €	5 000		
planning cost, % of total investment	10		
energy and water cost			
electricity costs - energy, €/kWh	0.1		
electricity costs - power (annual peak), €/kW	150		
fuel costs, €/kWh	0.05		
water costs, €/m³	3		
maintenance			
annual maintenance costs of solar system; % of investment	1		
annual maintenance costs of other components, % of investment	2		
capital cost parameters			
expected lifetime of solar system, a	20		
interest rate on investment of solar system, %	6		
expected life time of other components, a	15		
interest rate on investment of other components, %	6		
parameters related to environmental performance			
conversion factor for electricity, kWh electricity per kWh primary energy	0.36		
conversion factor for fossil fuel, kWh fossil fuel per kWh primary energy	0.95		
CO_2 emission from electricity generation, kg/kWh	0.8		
CO_2 emission from fuel (gas), kg/kWh	0.3		

A3 APPENDIX 3

The Solar Heating & Cooling Programme of the International Energy Agency (IEA)

The International Energy Agency (IEA) was formed in 1974 as an autonomous body within the Organisation for Economic Co-operation and Development (OECD). It carries out a co-operative programme on issues concerning energy, including joint research and development of new and improved energy technology.

The Solar Heating and Cooling (SHC) Programme was one of the first IEA research agreements to be established. Since 1976, its members have been collaborating to develop technology that use the energy of the sun to heat, cool, light and power buildings. The following 20 countries, as well as the European Commission, are members of this agreement: Australia, Austria, Belgium, Canada, Denmark, Finland, France, Germany, Italy, Japan, Mexico, the Netherlands, New Zealand, Norway, Portugal, Spain, Sweden, Switzerland, United Kingdom, United States.

The mission of the SHC Programme is: to facilitate an environmentally sustainable future through the greater use of solar design and technology.

Current Tasks of the IEA Solar Heating and Cooling Programme are:

Task 25: Solar-Assisted Air-Conditioning of Buildings

Task 27: Performance of Solar Facade Components

Task 28: Sustainable Solar Housing

Task 29: Solar Crop Drying

Task 31: Daylighting Buildings in the 21st Century

Task 32: Advanced Storage Concepts for Low Energy buildings

Task 33: Solar Heat for Industrial Processes

For more information: *http://www.iea-shc.org*

A4 APPENDIX 4

TASK 25 SOLAR-ASSISTED AIR-CONDITIONING OF BUILDINGS

Duration: June 1, 1999 - May 31, 2004

Objectives

The objective of Task 25 is to improve conditions for the market introduction of solar-assisted cooling systems in order to promote a reduction of primary energy consumption and electricity peak loads due to air-conditioning and thereby to develop an environmentally friendly method of air-conditioning of buildings. Therefore, the Task 25 focuses on:

- The definition of performance criteria of solar-assisted cooling systems considering energy, economic performance and environmental benefits;
- The identification and further development of promising solar-assisted cooling technology;
- An optimised integration of solar-assisted cooling systems into the building and the heating, ventilation and air-conditioning (HVAC) system focusing on an optimised primary energy saving - cost performance; and
- The creation of design tools and design concepts for architects, planners and civil engineers.

For more information about Task 25:

Website of the IEA Solar Heating & Cooling Programme:
http://www.iea-shc.org

Website of Task 25:
http://www.iea-shc-task25.org

Task Participants

Austria

Wolfgang Streicher
TU Graz
Inffeldgasse 25; A-8010 Graz

Michael Neuhäuser
arsenal research - Geschäftsfeld Erneuerbare Energie / Business Area Renewable Energies
A-1030 Vienna, Faradaygasse 3, Objekt 210

France

Jean-Yves Quinette (Leader of Subtask D - Solar-Assisted Cooling Demonstration Projects)
Tecsol
105, rue Alfred Kastler - Tecnosud - B.P. 434;
F-66004 Perpignan

Daniel Mugnier
Tecsol
105, rue Alfred Kastler - Tecnosud - B.P. 434;
F-66004 Perpignan

Rodolphe Morlot
CSTB
Boite postale 209
F-06904 Sophia Antipolis Cedex

Germany

Hans-Martin Henning (Operating Agent)
Fraunhofer Institute
for Solar Energy Systems ISE
Heidenhofstr. 2, D-79110 Freiburg

Uwe Franzke (Leader of Subtask B - Design Tools and Simulation Programs)
Institut für Luft- und Kältetechnik
(ILK Dresden)
Bertolt-Brecht-Allee 20
D-01309 Dresden

Carsten Hindenburg
Fraunhofer Institute
for Solar Energy Systems ISE
Heidenhofstr. 2, D-79110 Freiburg

Tim Selke
Fraunhofer Institute
for Solar Energy Systems ISE
Heidenhofstr. 2, D-79110 Freiburg

Mario Motta
Fraunhofer Institute
for Solar Energy Systems ISE
Heidenhofstr. 2, D-79110 Freiburg

Jan Albers
Institut für Erhaltung und Modernisierung von Bauwerken e.V. an der TU Berlin
Salzufer 14, 10587 Berlin

Michael Kaelcke
ZAE Bayern Abtl./Div.#1
Walther-Meissner-Str. 6, D-85748 Garching

Christoph Kren
ZAE Bayern Abtl./Div.#1
Walther-Meissner-Str. 6, D-85748 Garching

Greece

Constantinos A. Balaras
National Observatory of Athens,
Institute for Environmental Research & Sustainable Development,
Group Energy Conservation
I. Metaxa & Vas. Pavlou,
GR-15236 Palea Penteli

Israel

Gershon Grossman
Technion - Israel Institute of Technology
Haifa 32000

Italy

Federico Butera
Politecnico di Milano, Dipartimento di Scienze e Tecnologie dell'Ambiente Costruito,
Via Bonardi 3, I-20133 Milano

Marco Beccali
Università degli Studi di Palermo,
Dipartimento di Ricerche Energetiche e Ambientali (DREAM)
Viale delle Scienze - edificio 9,
I-90100 Palermo

Japan

Hideharu Yanagi
Mayekawa MFG.Co., Ltd., 2000,
Tatsuzawa Moriya-Machi,
Kitasoma-Gun, Ibaraki-Pref. 302-0118

Mexico

Isaac Pilatowsky (Leader of Subtask A - Survey of Solar-Assisted Cooling)
Universidad Nacional Autonoma de Mexico
Apdo. Postal #34. Temixco 62580, Morelos

Roberto Best
Universidad Nacional Autonoma de Mexico
Apdo. Postal #34. Temixco 62580, Morelos

Wilfrido Rivera
Universidad Nacional Autonoma de Mexico
Apdo. Postal #34. Temixco 62580, Morelos

Netherlands

Rien Rolloos
TNO Building and Construction Research
Schoemakerstraat 97,
P.O. Box 49, NL-2600 AA Delft

Daniel J. Naron
TNO Building and Construction Research - Sustainable Energy and Buildings
Schoemakerstraat 97,
P.O. Box 49, NL-2600 AA Delft

Gerdi R. M. Breembroek (Leader of Subtask C - Technology, Market Aspects and Environmental Benefits)
IEA Heat Pump Centre / Novem bv
P.O. Box 17, 6130 AA Sittard

Cees Machielsen †
TU Delft
Mekelweg 2, NL-2628 CD Delft

Portugal

Manuel Collares-Pereira
INETI-DER
Estrada do Paco do Lumiar
P-1649-036 Lisboa

João Farinha Mendes
INETI-DER
Estrada do Paco do Lumiar
P-1649-036 Lisboa

Maria Joaó Cavalhó
INETI-DER
Estrada do Paco do Lumiar
P-1649-036 Lisboa

Spain

Jordi Cadafalch Rabasa
Universitat Politècnica Catalunya
Colom, 11, E-08222 Terrassa (Barcelona)

Carlos David Pérez Segarra
Universitat Politècnica Catalunya
Colom, 11, E-08222 Terrassa (Barcelona)

Hans Schweiger
AIGUASOL Enginyería
Palau 4, 1er 1era, E-08002 BARCELONA

Laura Sisó Miró
AIGUASOL Enginyería
Palau 4, 1er 1era, E-08002 BARCELONA

SpringerArchitecture

Andrew Watts

Modern Constuction: Facades

2005. 200 pages. Numerous figures, partly in col.
Format: 21 x 29,7 cm
Hardcover **EUR 69,–**
Recommended retail price. Net price subject to local VAT.
ISBN-13 978-3-211-00638-2
Modern Construction Series

The Facades Handbook is a textbook for practitioners of architecture, as well as structural and environmental engineers who wish to broaden their study beyond the information provided in the "Walls" chapter of the Modern Construction Handbook. The six chapters examine facades from the standpoint of the primary material used in their construction, from metal to glass, concrete, masonry, plastics and timber. Each set of three double page spreads explains a specific form of construction which is accompanied by drawn and annotated details.
Throughout the book, built examples by high profile designers are used to illustrate specific principles. As is the case in the Modern Construction Handbook, the techniques described can be applied internationally.

P.O. Box 89, Sachsenplatz 4–6, 1201 Vienna, Austria, Fax +43.1.330 24 26, books@springer.at, **springer.at**
Haberstraße 7, 69126 Heidelberg, Germany, Fax +49.6221.345-4229, SDC-bookorder@springer.com, springer.com
P.O. Box 2485, Secaucus, NJ 07096-2485, USA, Fax +1.201.348-4505, service@springer-ny.com, springer.com
All errors and omissions excepted.

SpringerArchitecture

Andrew Watts

Modern Construction: Roofs

2005. 200 pages. Num. figs., partly in col. and 300 detailed drawings.
Format: 21 x 29,7 cm
Hardcover **EUR 69,–**
Recommended retail price. Net price subject to local VAT.
ISBN-13 978-3-211-24071-7
Modern Construction Series

The Roofs Handbook is a textbook for practitioners of architecture, as well as for structural and environmental engineers who wish to broaden their study. The six chapters examine roofs from the standpoint of the primary material used in their construction, from metal, glass, and concrete to timber, plastics and fabrics. Each set of three double page spreads explains a specific form of construction which is accompanied by drawn and annotated details.
Throughout the book, built examples by high profile designers are used to illustrate specific principles. As it is the case in the Modern Construction Handbook the techniques described can be applied internationally.

P.O. Box 89, Sachsenplatz 4–6, 1201 Vienna, Austria, Fax +43.1.330 24 26, books@springer.at, **springer.at**
Haberstraße 7, 69126 Heidelberg, Germany, Fax +49.6221.345-4229, SDC-bookorder@springer.com, springer.com
P.O. Box 2485, Secaucus, NJ 07096-2485, USA, Fax +1.201.348-4505, service@springer-ny.com, springer.com
All errors and omissions excepted.

Springer-Verlag
and the Environment

WE AT SPRINGER-VERLAG FIRMLY BELIEVE THAT AN international science publisher has a special obligation to the environment, and our corporate policies consistently reflect this conviction.

WE ALSO EXPECT OUR BUSINESS PARTNERS – PRINTERS, paper mills, packaging manufacturers, etc. – to commit themselves to using environmentally friendly materials and production processes.

THE PAPER IN THIS BOOK IS MADE FROM NO-CHLORINE pulp and is acid free, in conformance with international standards for paper permanency.